FOUNDATION OF QUANTUM GRAVITY-THE NUCLEUS-BLACK HOLE PARALLEL

BALUNGI FRANCIS

2015

Dedication+

This book is dedicated to my wife

Wanyana Ritah,

My lovely son Odhran Tumwebaze

My sisters,

Nantumbwe Florence and Kemirembe Prossy

Abbreviations

G- *the fundamental gravitational constant*

K_e-*the coloumb constant*

e-*charge on an electron*

h- *planck constant*

$\hbar = \frac{h}{2\pi}$-*reduced planck constant*

ε_o-*permitivity of free space*

μ_o-*permiability of free space*

c-*constant speed of light*

E-*electric field*

B-*magnetic fieldm- mass of a particle*

v-*velocity of a particle*

t-*time*

W-*work done/energy*

r-*radius of orbit*

n-*principle quantum number*

α-*coupling constant*

f-*frequency*

λ-wavelength

Contents

1.0. Introduction

Many books have been written about the theory of quantum gravity but in this book a different view of the theory is presented, in this book we try to study the theory from new assumptions which are far more different from models that have been used by scientists for centuries. Most physicists have clung to old models or complex mathematical scientific methods to explain phenomenon. They are trying to explain physics using the mathematics that was earlier used by Einstein, Richard P. Feyman and other scientists. The mathematics that was used by these physicists was complicated and has been difficult to understand which has led to students of physics in higher institutions to get discouraged about physics as a science.

Physics as a subject should be simple and elegant, trying to explain everything from one source, in other words trying to explain all of physics from one equation call it "the principle of least action". Imagine deducing the equations of gravity, quantum mechanics, electromagnetism, heat etc from one equation, wouldn't it be unique than holding about ten books about a different subject of physics each starting from its own source?

Physics is an entity and therefore requires only one subject to describe it fully and this subject is quantum gravity. A student of physics is therefore required to carry only one book for his entire physics life. In this book all physical theories have been explained and deduced from one equation implying that a simple equation can give rise to a complex field of physics from a somewhat simple rule or model of its calculation.

Many physicists do not understand this idea; they rather wait for a bomb to burst in their lab experiments or a miracle to happen in their calculations. There will be no miracle for the theory of quantum gravity, we can wait for a billion years but I tell you a miracle will not happen, see Einstein the most intelligent physicists waited for almost 35 year but no miracle happened in his paper the same applies to you, you will wait and no results will happen. Let us try to use our common sense to create something out of something. If the theory of gravity and quantum gravity are present we should only ask our selves can we combine the two and if combined is it possible for other theories to be derived from the combined one or what are the predictions of the theory? This is what this book is all about.

As a new student in this field of quantum gravity what questions will you ask your self if at all you are required to combine the quantum theory with the gravitational theory? Below I present to you the steps required for one to combine gravity with quantum mechanics. But first let us try to explain the two terms raised that is, gravity and quantum mechanics in relation to the theory of quantum gravity.

CHAPTER ONE

1.2. Gravity and Quantum mechanics

1.2.1. Gravity

By definition; to a person standing on the earth's surface, gravity is a force that pulls us towards the center of the earth. The force of gravity is realized when the phenomena being studied has a large mass. In other words the force of gravity is felt between two large bodies, that is, between the earth and the sun. Also the gravitational force is what keeps the earth in orbit around the sun and its strength increases as the distance between the two bodies' increases. For purposes of this book we disregard Einstein's General relativity theory/its mathematics and interpretations of gravity has gravitons. For purposes of this book we strongly reject any verbal and non verbal ideas about gravitons as these have lead to students who love the subject to be confused.

1.2.2. Quantum mechanics or Quantum theory

This is a mathematical description of subatomic particles and how they interact. It describes the strength of the electromagnetic force between two electrons by an exchange of quanta known as a photon. The quantum theory also explains the wave particle duality of matter and its interactions. For purposes of this book the langragian, laplace equations will not be used. The descriptions in this book are based on common sense and are not complicated in anyway.

To fully understand the theory of quantum gravity one needs to know the meaning of light. **What is light?** In a broad description, light is an electromagnetic wave made up of particles (photons) of discrete energy bundles like bullets shot from a gun. Light can also be bent when reflected towards a strong gravitational field. This definition is what is called a wave- particle duality.\

The light we have described above is the same light that is emitted from an atom whenever an electron in orbit around the nucleus jumps from one energy level to another. It is also the same light that is emitted by the sun. Whatever happens in the atom is the beginning to our development of the theory of quantum gravity, because like the earth's orbit around the sun it is the same mechanism by which the electron orbits the nucleus of an atom. The difference between the electron-nucleus system and the earth sun solar system is that in the atom the force that keeps an electron in orbit is the electromagnetic force while that of the earth's orbit is the gravitational pull. It has been said that, "the effects of gravity in an atom are so small and therefore negligible" but in this book they are introduced into an atom using a postulate that states; *two energies are possessed by an electron whenever it jumps from one energy level to another and that is, the energy of a photon and the potential gravitational energy possessed by an electron when moved from one state to another*. The two energies will never be equal since the speed of light of the emitted photon/electromagnetic wave will never be equal to the speed of an electron in orbit around the nucleus.

$$\frac{energy\ of\ a\ photon}{gravitational\ potential\ energy} = \frac{speed\ of\ light/photon}{speed\ of\ an\ electron}$$

By definition: the speed of light is related to the electric and magnetic fields by $c = \frac{E}{B}$, also the electric field is defined as the electric force per unit charge for an electron placed in vicinity of an electric field $E = \frac{F_e}{e}$, but the magnetic field in this case for a moving electron at a speed v is $B = \frac{F_b}{ve}$. The gravitational potential energy is the product of the gravitational force and the radius of orbit of an electron r $F_g\,r$. When the above definitions are taken into account then the energy of the emitted photon is related to the gravitational and electromagnetic force by

$$w_p = \frac{F_g F_e}{F_b} \qquad\qquad (1)$$

We have therefore hypothesized that the gravitational , the electric and magnetic force acting at a radius r of orbit of an electron around the nucleus of an atom are all relevant in the emission of a photon whenever an electron jumps from one energy level to another level.

12

1.2.3. The intensity of an electron in the electric field

The speed of a photon /light emitted from an atom due to the electric force is given by $c = \frac{F_e r^2}{n\hbar}$ and the angular momentum of an electron in orbit due to the gravitational force is $mvr = \frac{F_g r^2}{c}$.

But since the speed of light doesnot change in all situations or by substituting the speed of light form equation $c = \frac{F_e r^2}{n\hbar}$ into the equation of the angular momentum, then the angular momentum of an electron is given by

$$mvr = \frac{F_g}{F_e} n\hbar \qquad\qquad (2)$$

If we treat the electron to be moving at a speed of light in the magnetic field then the power of an electron will be given as a product the magnetic force on an electron and the speed of light as $p = F_b c$, we can generalize the power of an electron on a quantum scale by making F_g the subject from equation (2) above and substituting it in equation (1). Take note that in this case the energy of a photon will be given by $W_p = h\frac{c}{\lambda}$. The power of an electron in orbit around the nucleus of an atom wll be given by,

$$p = \frac{2\pi r^2 \lambda m v F_e{}^2}{nh^2}$$

since an electron in orbit around the nucleus exbhits deBrogile wave properties and also moves in a spherical shape then the two assumptions hold that , if the wavelength $\lambda = \frac{h}{mv}$ and the surface area of a sphere is $A = 4\pi r^2$ and the intensity is power per unit area $I = \frac{P}{A}$, then the intensity of an electron on a quantum scale will be given by,

$$I_n = \frac{E^2 e^2}{2nh} \qquad\qquad (3)$$

This implies that for each energy level the intensity of an electron is proportional to the square of its electric field.

However on a quantum gravity scale we can generate the intensity of a massive body in the gravitational field by taking an hypothesis that, if we treat the electromagnetic force Ee as an inverse square law we obtain the force on the particles in orbit as with the case for an electron in orbit around a nucleus at a distance r, then the electromagnetic force between two particles will be given by

$$Ee = \frac{n^2 \hbar c}{8\pi r^2}$$

For the Schwarzichild's radius $r = \frac{Gm}{c^2}$, the electromagnetic force is given as $Ee = \frac{n^2 \hbar c^5}{8\pi G^2 m^2}$, substituting this expression into equation3 **then the intensity on a quantum gravitational scale** will be given by

$$I = \frac{n^3 \hbar c^{10}}{256 \pi^3 G^4 m^4}$$

The intensity of an electromagnetic wave can also be deduced from eqn3 by assuming that at a level where the principle quantum number n is equal in magnitude to the fine structure constant or the electromagnetic coupling constant $\alpha = \frac{e^2}{4\pi\varepsilon_0 \hbar c}$ then the intensity of an electromagnetic wave in the electric field and the combined electric and magnetic field is given by

$$I = 2\varepsilon_0 E^2 c = \frac{EB}{\mu_0} \qquad\qquad (4)$$

This implies that, the properties of an electromagnetic wave will only be present whenever the principle quantum number is the electromagnetic coupling constant or fine structure constant of value $n = \alpha = \frac{1}{137}$.

CHAPTER TWO

2.1. Black hole radiations/ Hawking radiations

By definition as of Wikipedia, a black hole is a mathematically defined region of space time exhibiting such a strong gravitational pull that no particle or electromagnetic radiation can escape from it. Many theories have been created to explain the properties of the black hole but the theory created here is far more different from the other theories although it may give the same results. Using a quite different approach towards solving a problem is efficient since it comes with it new predictions in the process which could have been hidden in other approaches. Below we try to present adhoc proofs-laws that may be of help in building our theory about black holes, note; these proofs can be derived mathematically from equations 1 up to 4 above but for purposes of simplicity they have been listed here below, however their derivations will be given in the last chapters of this book.

The laws or equations:

1) It is well known that the electric field is force per unit charge but here a generalized equation for an electric field created by an electron exhibiting wave properties in the nucleus of an atom in the gravitational field on a quantum scale is given by

$$E = \frac{1}{r}\sqrt{\frac{Gm^3 f}{2\hbar\varepsilon_0}} \qquad (5)$$

Then the electric force in this case will be formulated as

$$F_1 = \frac{e}{r}\sqrt{\frac{Gm^3 f}{2\hbar\varepsilon_0}} \qquad (6)$$

2) The surface area at a radius r of orbit of an electron of mass m around the nucleus of an atom in a wave like manner is given by

$$surface\ area(A) = \frac{\lambda\mu_0 e^2}{m} \qquad (7)$$

3) The time taken by the magnetic field B of an electron to pass through a given surface is

$$time(t) = \frac{\lambda\varepsilon_0 AB}{e} \qquad (8)$$

Note: the above expression is the same as Faraday's induction law.

4) The gravitational force acting on all matter in the universe or the modified gravitational force is given as

$$F_2 = \left(\frac{Gm^3}{r^2}\right)\left(\frac{e}{2B\lambda\hbar\varepsilon_0}\right) \qquad (9)$$

The above formulas are important in deriving the formula for the temperature, entropy and the time taken by a black hole to evaporate as shown below;

17

2.1.1 Temperature of a black hole

It is known that the kinetic energy KE of molecules in the Boltzmann hypothesis is related to the temperature of the body in question in this case a black hole (in relation to the black body) by $KE = \varphi T$ where φ is Boltzmann's constant. The formula for the kinetic energy can be derived by using a hypothesis that the electromagnetic force – coulombs force is equal to eqn6 as

$$\frac{ke^2}{r^2} = \frac{e}{r}\sqrt{\frac{Gm^3 f}{2\hbar\varepsilon_o}}$$

On squaring both sides of the equation, cancelling like terms and taking into account that the frequency of an electron is $f = \frac{v}{\lambda}$, then the kinetic energy of an electron inside the black hole is given by

$$KE = \frac{\lambda\mu_o e^2}{A}\frac{c^3\hbar}{8\pi Gm^2}$$

Since the surface area is given as from eqaution7 then **the kinetic energy of molecules** or particles (for an ideal gas) within the black hole will be given by

$$KE = \frac{c^3\hbar}{8\pi Gm} = T\varphi \qquad (10)$$

Then from Boltzmann's relationship the temperature of the black hole is formulated as

$$T = \frac{c^3 \hbar}{8\pi Gm\varphi}$$
(11)

2.1.2 The entropy of the black hole

By definition entropy is a measure of disorder. To solve the entropy of black holes we shall consider a very complex argument about the entropy in question. We assume that the modified gravitational force given by equation 9 is identical to the modified electric field given by equation6 as, $\left(\frac{Gm^3}{r^2}\right)\left(\frac{e}{2B\lambda\hbar\varepsilon_0}\right) \equiv \frac{e}{r}\sqrt{\frac{Gm^3 f}{2\hbar\varepsilon_0}}$ in otherwise the two forces are equal but opposite. Then squaring both sides of the equation and multiplying through by Gc^5 one obtains a new relation of forces on both sides given as

$$\frac{tc^7}{16\pi G^2 m} = \frac{Ac^6}{32\pi rmG^2}$$

Both the left and right hand side represent a force. From the left hand side t is the expression of time given by $t = \frac{\hbar e^2}{2m^3 c^2 G\varepsilon_0}$. Note: the left hand side force is the pull of matter inside the black hole while the right hand side force is the force acting on particles or matter at the surface of the black hole.

Since the heat is the product of the force on a particle and the distance r from the centre of the black hole, then using the force

19

on the right hand side of the above equation the heat will be given by

$$Q = \frac{Ac^6}{32\pi mG^2}$$

Remember the temperature of the black hole is also known from equation9 and by definition the entropy of the system is the change in heat per unit temperature $\frac{Q}{T}$, then the entropy of the black hole will be given by

$$S = \frac{A\varphi c^3}{4Gh} \qquad\qquad (12)$$

This implies that the entropy of a black hole is proportional to its surface area.

2.1.3   The time taken by a black hole to evaporate

Assuming that particles that formed a black hole are moving away or are separating from it after a given time of its existence, if we measure the relative speed of these particles in relation to the energy they carry we obtain a relation ship given by

$$\frac{v^2}{c^2} = \frac{8\pi G}{c^2}\left(\frac{W}{8\pi r}\right) \qquad\qquad (13)$$

Where v is the velocity of these particles as measured relative to the speed of light c and W is the energy carried by the particles as they move away from the centre of the black hole at a distance r.

If we let the force causing the particles to separate from the black hole be given as $\dfrac{Gm^3 e}{2r\lambda B\hbar\varepsilon_0 c}\dfrac{v}{}$, then the energy of these particles will be given by

$$W = \frac{Gm^3 e}{2r\lambda B\hbar\varepsilon_o c}\frac{v}{}$$

Substituting this in equation11, we obtain a relation ship of time as given by the law 3 of equation 8 as

$$t = \frac{v^2}{c^2}\left(\frac{\pi G^2 m^3}{\hbar c^4}\right)$$

The velocity of the particles in the astronomical lab will be measured as v= 4.193E6 m/s and since the speed of light is a constant then the time taken by a black hole to evaporate is given by

$$t = \frac{5120\pi G^2 m^3}{\hbar c^4} \qquad\qquad (14)$$

CHAPTER THREE

3.0. A general and simple classical unified field theory

The development of a unified field theory of physics over so many years has involved the use of complicated mathematics coupled with assumptions and imaginations that have made it impossible for the creation of a simple and beautiful unified field equation. The mathematics for unification should be concise and elegant.

Attempts to unify gravitation and electromagnetism came a long way with the realm of Albert Einstein in what was called a classical unified field theory. However such attempts were abandoned with the development of modern theories of physics i.e. The General Relativity theory and the Quantum theory. These two theories are elegant and predictive in their domain of applicability.

A classical unification theory should reproduce these two theories and also be able to explain it. Today we know of the standard model and general relativity and how they are successful, but the unification of the two theories has also proved to be difficult. In my conscience these should be fruits of a classical unified field theory. In other words a classical unification incorporates them as one that is they are supposed to appear in the unification equation. Therefore the development of modern physics acts as a check for classical physics in any field of physics concerned with unification.

The objective of this paper is to develop a unified field theory uniting gravity and electromagnetism using classical physics and non- quantum approaches.

3.1.1The unified electromagnetic gravitational oscillations

The gravitational potential field (gravity) is caused by the presence of any mass in space. Like gravity, charge is a fundamental property of the universe. Just as mass causes a gravity field, charge causes an electromagnetic field. There are no known methods of insulation of gravitational field in modern science. It is impossible to imagine space and gravitation separately. Gravitation exists everywhere where there is some space. The gravitational field created by all masses of our metagalaxy is the aether in which cosmic objects moves and electromagnetic oscillations are propagated. The space surrounds us since the whole matter carries gravitational charge of only one sign.

A dipole antenna is used to produce overlapping electric and magnetic fields E and B respectively. Such an antenna is so far the simplest practical antenna from a theoretical point of view, and therefore is the basis on which this study is conducted. The current amplitude I for such an antenna decreases uniformly from maximum at the center to zero at the ends.

The work done to move the electron at a speed v(t) up and down the antenna is the gravitational potential energy W_G and is proportional to the velocity v(t). The electric potential energy W_E of the electron at a distance R from the antenna are proportional to their angular momentum L. $W_G = k_1 v (t)$ and $W_E = k_2 L$, if the proportionalities are true then also the following should be true WG $W_E = \beta$ L (t) v (t) where $k_1 k_2 = K^2 = \beta$ is the force required to accelerate the electrons.

Differentiating W_G with respect to height h moved by a charge up the antenna, and W_E with respect to the charge q on the electron i.e. $dW_G /dh = mg$, $dW_E/dq = ER$. Then the product of the acceleration due to gravity g and the electric field E is

Eg = (βL/mRt)div   **1**

∇ **= div = d/dq**

When a charge is moved up and down the antenna by the oscillator in the y direction a changing magnetic field is produced in the z direction leading to a changing gravitational field. Then the acceleration due to gravity is of magnitude g= vcΩ, Ω is the curvature of space in regions near the antenna where a charge experiences the electric and magnetic field. And c is the speed of the electromagnetic wave transmitted from the antenna. Substituting for g in equation 1 gives

$$\nabla\beta = d\,\beta\,/dr = f_e\Phi \qquad\qquad 2$$

This is the rate of change of force with the radius of a spherical wave front r=cT, T is the time- dimension and then $r^2 = x^2 + y^2 + z^2 + T^2$. Therefore $\nabla\beta$ depends on the electromagnetic force f_e=Eq and the curvature $\Phi=\sqrt{BI/W}$. where W is the kinetic energy of the electron and I is the current.

According to R.L. Collins, (1997-2006) Energy is exchanged between neutral masses, via a long range electromagnetic force, and that this exchange of energy reproduces the effects of gravity. If β is taken as the gravitational force acting perpendicular through the center of mass of the particle then the potential gravitational gradient will be given by

$$\nabla^2 W_G = f_e\Phi \qquad\qquad 3$$

The electromagnetic wave transmitted from the antenna will move along the x-axis. Since this wave is moving through a vaccum then there can be no conventional currents, but there is a displacement current, hence I =0 and Φ=0. Then the condition $\nabla^2 W_G = 0$ is fulfilled. The accelerated motion of electrons up and down a straight-rod dipole antenna therefore produces electromagnetic waves that satisfy the above equation. The

25

electric, magnetic and gravitational fields are periodic in both time and space.

3.1.2 Discussion and conclusions

Change of both electric and gravitational field results in the creation of a magnetic field in the region of space-time which has a dual electrogravitational nature. The amplitude of electromagnetic and magnetogravitational constituents of the unified electromagnetic gravitational oscillations depends on field potential of opposite nature. The electromagnetic constituent is determined by gravitational potential and the magnetogravitational one is determined by electric potential. Transference of gravitational masses of matter in electrogravitational field-aether causes the creation of the proper magnetic field. (V.Ya.Kosyev, 2000).In the quantum realm, the gravitational force is so weak that it is difficult to observe quantum effects caused by gravity. However Nesvizhevsky and collaborators have reported an experiment in which they observed quantum effects of gravity on the behaviour of ultracold neutrons (UCNs). These neutrons have kinetic energies so low that they can be trapped by gravity above a reflecting surface. (Thomas J. Bowles, 2002)

CHAPTER FOUR

4.1. Foundations of quantum gravity theory-The Nucleus-Black Hole parallel

In the early 20th century, Ernest Rutherford experiments established that atoms consisted of a diffuse cloud of negatively charged electrons surrounding a small, dense, positively charged nucleus. Given this experimental data, it was quite natural for Rutherford to consider a planetary model for the atom, the Rutherford model of 1911, with electrons orbiting a sun-like nucleus. This model was a difficulty. The laws of classical mechanics predict that the electron will release electromagnetic radiation as it orbits a nucleus. Because the electron would be losing energy, it would gradually spiral inwards and collapse into the nucleus. This is a disaster, because it predicts that all matter is unstable.

To overcome this difficulty, Niels Bohr proposed, in 1913, what is now called the Bohr model of the atom. The model's key success lay in explaining the Rydberg formula for the spectral emission lines of atomic hydrogen. Not only did the Bohr model explain the reason for the structure of the Rydberg formula, but it provided a justification for its empirical results in terms of fundamental physical constants.

This paper looks at the model in a very different way than that of Bohr. The fact that all accelerated particles do emit electromagnetic radiations is taken into account and therefore the acceptance for the unstableness of all matter is considered in due respect. In fact Bohr's ideas never required classical mechanics simply because it could not conform to the experimental observations of the spectrum of the Hydrogen atom that were obtained by Rydberg using his formula.

To merge gravity with Planck's quantum theory by then was also a problem at hand and therefore Bohr had to forego the problem by introducing in his theory adhoc postulates, and this could have been the reason why Einstein found problems in merging gravity with electromagnetism in what is called "The Grand unified field theory", of which he had to question the problem with the quantum theory and therefore request for a complete quantum theory. From Bohr's model many theories have been formed each building from the ideas of the model, but a certain point is reached where the theories can not conform well to the known laws of nature and therefore regarded as failures, which of course in their judgments is true. The problem is seen to come from exactly the roots of quantum mechanics.

The aim of this paper is therefore to produce a generalized theory of atomic structure that incorporates in it gravity and quantum mechanics. In other words a theory that takes the laws of classical mechanics into consideration.

4.1.1Methodology

The Hydrogen atom exists in certain stationary states of discrete energies. The acceleration due to gravity of an electron in orbit around the nucleus will cause the atom to emit radiations (radiate energy) and thus make the atom unstable. The acceleration (g) falls off with time t provided the radius of orbit of the electron R is a constant thus the acceleration due to gravity is given by;

$$g= R/\Delta t^2 \qquad (1)$$

The rate of change of energy P radiated as a result of the above acceleration will depend on the constants c (speed of light) and G (universal gravitational constant), hence;

$$P= c^5/G \qquad (2)$$

The power and time must be re- quantized in units of $\hbar = h/2\pi$ where h is Planck constant, hence

$$P\Delta t = n^2\hbar \qquad (3)$$

Where n= 1,2,3…….. is the principle quantum number.

But the total energy of the atom in the various energy states is W= $-ke^2/R$ where k is the Coulomb constant and e is the elementary charge. Since Δt^2 is known from Eqn1 and P from Eqn2 then using Eqn3 the radius is given by

29

$$R = n^2 Gg\, \hbar\, /c^5 \qquad (4)$$

From which the total energy is given by,

$$W = -\, ke^2 c^5 /\, n^2 Gg\, \hbar \qquad (5)$$

From the Bohr-Einstein frequency (f) condition, applied to a transition from a level with $n = n_i$ to a level with $n = n_f$, The energy of a photon emitted by a hydrogen atom is given by the difference of two hydrogen energy levels

$$hf = E_i - E_f$$

Finally we get since frequency $f = c/\lambda$, where λ is the wavelength

$$1/\lambda = [ke^2 c^4 /\, 2\pi G\hbar^2][1/g][1/\, n_f^{\,2} - 1/\, n_i^2] \qquad (6)$$

The equation obtained above shows some how a great significance of gravity in the quantum theory. So far it states that regardless of the levels in the transitions of an atom the acceleration due to gravity of the particles in the atom do greatly affect the nature of its spectrum.

4.1.2   Results

The quantity $[ke^2 c^4 /\, 2\pi G\hbar^2]$ is the inverse of the square of time t and therefore

$1/t^2 = [ke^2 c^4 /\, 2\pi G\hbar^2]$, from which the time is obtained as $t = 1.58873 \times 10^{-42}$ s.

Comparing Eq6 with Bohr's model, here we shall equate the Rydberg constant $[k^2e^4m/4\pi c\hbar^3]$, where m is the mass of the particle, to the constant $[ke^2c^4/\ 2\pi G\hbar^2][1/g]$. Doing this generates an acceleration given by $g_a= [\ 2\hbar c^5/ke^2\ Gm]$ from which we obtain a general equation of forces given by $[8\pi G/c^4][gm][ke^2\ /R^2] = 16\pi\hbar c/R^2$.where [gm] is the gravitational force and $[ke^2\ /R^2]$ is the electromagnetic force.

At the Schwarz child's radius $R=Gm/c^2$ the acceleration is $g_b = c^4/Gm$ which gives an equation for the spectrum as $1/\ \lambda = [/\ 2\pi\ a_0][1/\ n_f^{\ 2}-1/\ n_i^2]$ where a_0 is the first Bohr radius $[\hbar^2/\ mke^2] = 5.28 \times 10^{-11}m$.

The interesting part of it is that the ratio $g_b\ /\ g_a =[ke^2/2\hbar c]$ the fine structure constant. This result therefore explains the fine structure shown by the Hydrogen spectrum and thus suggests that an electron describes an elliptical orbit. Now using the acceleration g_a, the radius a_0 and the mass m the energy W of a particle will be given by $W =g_aa_0m =\beta/m$, where β is a constant given by $[\ 2\hbar^3c^5/k^2e^4G] =1.64367 \times 10^6Jkg$. For two different masses m and m_0 we have the equation for the product of the masses as, $mm_0 = [\ 2\hbar^3c^3/k^2e^4G]$.

4.1.3   Discussion

The results obtained give out a clear image for the description of the atomicity of both large and small particles. Firstly the time t obtained is the is the earliest period of time in the history of the universe from zero to approximately 10^{-43} seconds , during which

31

quantum effects of gravity were significant. At this period all the fundamental forces of physics were unified. The state of the universe during this epoch was unstable, tending to evolve and giving rise to the familiar manifestations of the fundamental forces through a process known as breaking. Symmetry breaking quickly led to the era of cosmic inflation, the Inflationary epoch, during which the universe greatly expanded in scale over a very short period of time.

Secondly, the accelerations g_a and g_b led to different spectrums of the Hydrogen atom. Where g_a produces the Rydberg equation for the spectrum of the hydrogen atom, that is incorporated in it the Rydberg constant $[k^2e^4m/4\pi c\hbar^3]$, the acceleration g_b produces a different equation which instead of a Rydberg constant, it has the inverse of the first Bohr radius a_o. These differences in the spectrum of the hydrogen atom with the former producing a single line and the latter two or more lines of the spectrum close together, imply that the electron moves in an elliptical orbit as those of the planets in orbit around the sun, hence the ratio of the accelerations g_b / g_a will generate a fine structure constant describing the closeness of the spectrum lines produced by the hydrogen atom. Finally the energy obtained using the acceleration g_a and the first Bohr radius a_o has an impact on the way we express the energy of large and small particles. For example a body of one kilogram mass (1Kg), will have an energy of 1 .64367 $\times 10^6$Joules (1 .64367 $\times 10^6$J) which is a very high energy. This energy is independent of the speed of a body

32

or particle in question, and thus gives the energy to a particle regardless of it's speed. We very well know that the speed of light is a constant and therefore doesn't change and that with relativity such a body of 1kg will have energy approximately 10^{16}J which is lager than the first one by 10^{10}. For smaller particles say an electron we have from the result equation the energy as 1.80425×10^{36}J, but for relativity it is $\sim 10^{-15}$J. These differences in energies imply that without knowing the speed of the particle we can obtain it's energy depending on it's mass since some particles tend to move at a speed greater than that of light.

4.1.4 Conclusion

In conclusion the results produced successfully indicate that without gravity, quantum mechanics can not survive and without quantum mechanics, gravity cannot survive. Therefore the two theories are needed to explain the atomic universe fully.

CHAPTER FIVE

5.1. The simplification of the nature and structure of particle physics

Mass-Energy; 1kg is equivalent to 5.6096×10^{26}GeV, Time; 1year $= 3.156 \times 10^{7}$s

Physicists have argued out that the more elegant and symmetrical the theory is, the more it is beautiful. The elegancy of any physical theory is suspected at a level to which it holds well with other theories , that is ,the capability of the theory to conform with the well known laws of nature at all levels.

In this paper we examine the mechanism through which quantum mechanics becomes comparable with gravity and the scale to which this occurs. At the Planck scale all interactions (the weak interaction, strong interaction and electromagnetism) are assumed to merge into a single interaction that alone occurs at very high energies of about 1TeV. The equations that do describe this phenomenon are not yet found and therefore require one's deep effort to capture the reality of this entire puzzle.

To capture interest in these interactions we need to know first, their strength and second the range in which they occur. The strength defines the coupling constants and the range defines the attractions, on the other hand the coupling constant determines the strength of any interaction and therefore is a number in a sense that it is a dimensionless constant. A coupling constant is a very important quantity in dynamics, for example, in the motion of a large lump of magnetized iron, the magnetic forces are more important than the gravitational forces because of the relative magnitudes of the coupling constants.

The Standard Model is a theory of three fundamental forces — electromagnetism, weak interactions and strong interactions; however, these three forces are not tied together. Howard Georgi and Sheldon Glashow discovered that the Standard Model particles can arise from a single interaction, known as a grand unified theory. Grand unified theories predict relationships between otherwise unrelated constants of nature in the Standard Model. Gauge coupling unification is the prediction from grand unified theories for the relative strengths of the electromagnetic, weak and strong forces and this prediction was verified at LEP in 1991 for supersymmetric theories.

In particle physics, supersymmetry (often abbreviated SUSY) is a novel symmetry that relates elementary particles of one spin to another particle that differs by half a unit of spin and are known as superpartners. Since the particles of the Standard Model do not have this property, supersymmetry must be a broken symmetry allowing the 'sparticles' to be heavy.

One of the main motivations for SUSY comes from the quadratically divergent contributions to the Higgs mass squared. The quantum mechanical interactions of the Higgs boson causes a large renormalization of the Higgs mass and unless there is an accidental cancellation, the natural size of the Higgs mass is the highest scale possible. This problem is known as the hierarchy problem. Supersymmetry reduces the size of the quantum corrections by having automatic cancellations between fermionic and bosonic Higgs interactions. If supersymmetry is restored at the weak scale, then the Higgs mass is related to supersymmetry breaking which can be induced from small non-perturbative effects explaining the vastly different scales in the weak interactions and gravitational interactions. The failure of experiments to discover either supersymmetric partners or extra spatial dimensions, as of 2006, has encouraged loop quantum gravity researchers.

5.2. Materials and methods

5.2.1 The determination of the strength of the forces

We assume a model that explains everything on the length scales, the best scale so far we are familiar with is the Planck length scale, however in this model we don't associate our selves in knowing this scale and there fore develop new scales that alone are combined together to lead to some observable phenomenon describing the forces involved in the interactions. The equation describing the model is developed and given by;

$$(v^2/c^2+n^2\beta_{Qo})=8\pi\beta_{gEo} \qquad (1)$$

Where β_{Qo} is a length ratio given by l_Q/l_o, in this case $l_Q = \hbar c/W$, $\hbar$ is Dirac constant, c is the speed of light and W is the energy. Also $\beta_{gEo} = l_{gE}/l_o$ where $l_{gE} = 8\pi GM_{gE}^2/W$, G is the universal gravitational constant, M_{gE} is the mass of a particle in the combined fields given by P_{gE}/c where $P_{gE} = Gm^2ke^2/R^2c^2$, is the momentum for an elementary particle of mass m and an elementary charge e, k is the coulomb constant and R is the distance between any two particles. The equation here addresses the problems in form of length scales simply because it is at these scales that quantum mechanics seem to be comparable to gravity. The momentum P_{gE} is a momentum of a particle experiencing the strength of the electromagnetic fields and gravity. The strength is determined by a very small coupling constant as we shall see later. The smaller the distance between elementary particles, the higher the momentum and vice versa is true.

The exchange of photons between an electron and a proton in an atom is explained by Quantum Electrodynamics (QED), with a coupling constant determining the strength of the electromagnetic force. The equation of the interaction responsible for QED on the length scale, which is the Compton length, is given by the equation

$$\sum \psi^2_{fi}t_i=2\pi\beta_{RCE} \qquad (2)$$

The expression $\sum \psi^2_{fi} t_i$ is the force changer where $\psi_{fi} = f_i R^2 / ke^2$ and $t = \{f_i^2 ke^2 / R^2\} / F_n^3$,

β_{RCE} remains a constant given by l_c / l_{RE} (l_c is the Compton length $\hbar/mc$ and $l_{RE} = ke^2/mR^2c^2$).

On multiplying both sides of Eqn1 by a quantity $\sum \psi^2_{fi} t_i$ we obtain,

$$(v^2/c^2 + n^2 \beta_{Qo}) \sum \psi^2_{fi} t_i = 8\pi \beta_{gEo} \sum \psi^2_{fi} t_i$$

We then examine the condition for which β_{Qo} will be a maximum and minimum. It is found out from relativity that β_{Qo} is maximum when the lorentz factor $\gamma = (1 - v^2/c^2)^{-1/2}$ is very small that is,

$\gamma = 1/n\sqrt{\beta_{Qo}}$ or when the velocity $v = c \sqrt{(\sum \psi^2_{fi} t_i - n^2 \beta_{QO})}$

We hence obtain a general interaction equation as,

$\sum f_y^3 \psi^2_{fi} t_i = 2\pi \beta_{RCE} \sum F_n^3 / \xi \beta_{gEo}, n = 1,2,3 \ldots \ldots (3)$

The following conditions are then taken into account

1) For $l_o = l_x = mc^2/Fp$, $\beta_{gEo} = \beta_{gEx} = l_{gE}/l_x$.

2) For $l_o = l_c$, $\beta_{gEo} = \beta_{gEQ} = l_{gE}/l_c$, and

3) For $l_o = l_s = Gm/c^2$, $\beta_{gEo} = \beta_{gEs} = l_{gE}/l_s$, which gives

$$\sum f_y{}^3 \psi^2{}_{fi} t_i = F_1{}^3 + F_2{}^3 + F_3{}^3 = 2\pi\beta_{RCE} (F_p{}^3 / 8\pi\beta_{gEx} + F_p{}^3 / 256\pi {}^3\beta_{gEQ} + 2F_B{}^3 / \pi \beta_{gEo}) \quad (4)$$

Where $F_p = c^4/G$ is the Planck unit force and $F_B{}^3 = m^2 c^3 / \hbar$ is the force required for strong and weak interactions to take place.

Again setting a condition,

For for $l_o = l_z = ke^2 / mc^2$, $\beta_{gEo} = \beta_{gEz} = l_{gE} / l_z$.

$$\sum f_y{}^3 \psi^2{}_{fi} t_i = F_4{}^3 = 2\pi\beta_{RCE} (F_z{}^3 / 32\pi^3 \beta_{gEx})$$

Where $F_z{}^3 = m^2 c^4 / ke^2$,

Also for $l_o = l_N = \hbar^2 m^3 G^2 / k^3 e^6$, $\beta_{gEo} = \beta_{gEN} = l_{gE} / l_N$, we obtain,

$$\sum f_y{}^3 \psi^2{}_{fi} t_i = F_5{}^3 = 2\pi\beta_{RCE} (F_z{}^3 / 2\pi^3 \beta_{gEN}) \quad (5)$$

Measuring the value of the strong, weak and electromagnetic coupling constants gives us away through which we can determine supersymmetric levels. From supersymmetry and grand unification of elementary particles the couplings agree to 1%. The relationships of the sum of the cubes of the forces to each individual cube of the force, and that of the sum of the square of masses with each known mass squared casts much information about the masses and couplings of the supersymmetric particles as shown below, when Eqn4 is divided through respectively by the cubes of the forces $F_1{}^3$, $F_2{}^3$ and $F_3{}^3$ the following equations are obtained,

$$\sum F_n{}^3 / F_1{}^3 = 1 + 16\alpha_g{}^3 + 1/32\pi^2 \alpha_g \quad (6)$$

$$\sum F_n{}^3/F_2{}^3 = 1 + 32\pi^2\alpha_g + 512\pi^2\alpha_g{}^4 \qquad (7)$$

$$\sum F_n{}^3/F_3{}^3 = 1 + 1/6\alpha_g{}^3 + 1/512\pi^2\alpha_g{}^4 \qquad (8)$$

$$\sum F_n{}^3/F_4{}^3 = 1 + \beta^2(4\pi^{2+}1/8\alpha_g) + 64\pi^2\alpha_s{}^2\alpha_g \qquad (9)$$

Where $\beta = ke^2/Gm^2$ is the ratio of the fine structure constant α_s to the gravitational coupling constant α_g, given respectively as $\alpha_g = Gm^2/\hbar c$ and $\alpha_s = ke^2/\hbar c$.

Now equating $F_4 = F_5$, $F_5 = F_3$, $F_5 = F_1$ we obtain; m_1, m_2 , m_3 and m_4 respectively, Adding the squares of the masses we obtain,

$$\sum m_n{}^2 = m_1{}^2 + m_2{}^2 + m_3{}^2 + m_4{}^2 \qquad (10)$$

Which gives the sum per unit mass as,

$$\sum m_n{}^2/m_1{}^2 = 1 + 16\pi^2\alpha_s{}^4 + (8\pi/\alpha_s)^{\frac{1}{2}} + 4/(128\ \alpha_s{}^4)^{1/5} \qquad (11)$$

$$\sum m_n{}^2/m_2{}^2 = 1 + 1/16\pi^2\alpha_s{}^4 + (1/8\pi\ \alpha_s{}^9)^{\frac{1}{2}} + (1/4\pi^2)(128\ \alpha_s{}^{24}) \qquad (12)$$

The equations generated so far give a basis for the nature and type of supersymmetry exhibited by a particle experiencing forces at both the Planck and grand unified scales. It is thus shown here that the electromagnetic coupling constant is a result of mathematically summing the squares of the masses generated and then dividing through by the square of the mass in the summation while the gravitational coupling constant is the result of summing the cubes of the forces and then dividing through by the cube of the force in the sum. This idea at its best is taken to be the basis for symmetric theories as we shall see in the results obtained.

5.3. Results

5.3.1 The unification of coupling calculations

At equal forces that is $F_1 = F_2 = F_3 = F_p$ the mass $M_p = (\hbar c /8\pi G)^{1/2} = 2.1765 \times 10^{-8}$ kg, is obtained which is the Planck mass for which the Schwarzschild radius is equal to the Compton length divided by π. When Eq4 is divided through by $F_1{}^6$ and $F_2{}^6$ we obtain equations of the form;

$$\sum F_n{}^3/F_1{}^6 = \Omega/F_p{}^3 \qquad (13)$$

$$\sum F_n{}^3/F_2{}^6 = \text{\euro}/F_p{}^3 \qquad (14)$$

Where, $\Omega = 4m^2/m_p{}^2 + 1/\pi + m^8/32\pi^2 m_p{}^8,$

$\text{\euro} = m^6/8\pi m_p{}^6 + 16\pi m^4/m_p{}^4 + 2m^{12}/\pi m_p{}^{12}$

The mass relations equations obtained above indicate the scale at which gravity may be strong and weak. Obtaining these results on the Planck force and mass scale is evidence for the existence of the theory of quantum gravity. The values Ω and $\text{\euro}$ represent a series equation defined by increasing powers in the mass ratio (m/m_p). The mass m is assigned to any particle and the mass m_p is assigned to the Planck scale defining quantum gravity.

The unit of energy is M_Pc^2; the unit of electric charge is $\sqrt{hc/k}$, where k is coulomb constant and so forth. On the other hand, one cannot form a pure number from these three physical constants. Thus one might hope that in a physical theory where $\hbar$, c, and G were all profoundly incorporated, all physical quantities could be expressed in natural units as pure numbers. Within its domain, this paper has achieved it for example, imagining that there were just two quark species with vanishing masses. Then from the two integers 3 (colors) and 2 (flavors), $\hbar$, and c (without mass parameters), the spectrum of hadrons with mass ratios and other properties close to those observed in reality, emerges by through calculation (Ω and $\text{\texteuro}$) as indicated from Eqn13 and Eq14 shown above. The overall unit of mass is indeterminate, but this ambiguity has no significance within the theory itself. The results obtained show an ideal Planckian theory that alone does not contain any pure numbers as parameters. Thus, for example, the value $m_e/m_p=10^{-22}$ of the electron mass in Planck units is obtained from a dynamical calculation. This ideal might be overly ambitious, yet it seems reasonable to hope that significant constraints among physical observables will emerge from the inner requirements of a quantum theory which consistently incorporates gravity. The model therefore provides; first, the unification of couplings calculation. second, it points to a symmetry breaking scale remarkably close to the Planck scale (though apparently smaller by 10^{-2} to 10^{-3}), so there are pure numbers with much more 'reasonable' values than 10^{-22} to shoot

42

for. Third, it shows quite concretely how very large scale factors can be controlled by modest ratios of coupling strength, due to the logarithmic nature of the running of couplings (so that 10^{-22} may not be so 'unreasonable' after all).

While the above result is based on the study of the strength of the gravitational force, we now look for ways in which we can examine the strength of the electromagnetic force depending on the mass. This is done by dividing the sum of the squares of the masses (Eqn10) by the fourth power of the individual masses hence,

$$\sum m_n^2 / m_2^{4=} \omega / m_G^2 \qquad (15)$$

$$\sum m_n^2 / m_E^{4=} \lambda / m_p^2. \qquad (16)$$

Where $\omega = 1/ 16\pi^4 \alpha_s^9 + 1 /4\pi^2 \alpha_s^5 + 1/ (512\pi^8 \, \alpha_s^{19})^{1/2}$

$\lambda = 128\pi^3 \alpha_s^6 + 8192\pi^5 \alpha_s^{10} + 128\sqrt{\pi^3}\alpha_s^{11} + 128(\pi^{15}\alpha_s^{26})^{1/5}$

$m_E[(1/ 8\pi \, Ke^2)(\hbar^3 c^3 /G)^{1/2}]$ is the mass obtained when $F_4^3 = F_3^3$, and

$M_G = (Ke^2 / G)^{1/2}$ is the mass obtained when the electromagnetic force is equal to the gravitational force.

It can now be theorized that the strength of the electromagnetic force is determined by Eqn15 and 16 at which a series power equation in the fine structure constant defined by ω and λ is a constant.

5.3.2 The length scales at which the masses predicted by the standard model survive

The mass of the W and Z bosons (M_W, M_Z),Higgs particle (M_H) and the mass scale at the grand unification (M_{GUT}) are generated. We multiply a coupling constant μ with the force F_3^3, of which we equate to F_4^3 that is;

$$\mu \, F_3^{\,3} = F_4^{\,3}$$

From which

$$\mu = R_B^{\,2} / R_o^{\,2,}$$

Where R_o is the length scale determined experimentally and

$R_B = (8 \, \pi G k e^2 / c^4)^{\frac{1}{2}} = 6.9101 \times 10^{-36}$ m , which is greater than the Planck length.

So the equation that produces the different masses at R_o will be given by the square of the mass as,

$$M^2 = m_p^{\,2} / 8 \, \pi \mu \, \alpha_s^{\,2}$$

Where $\alpha_s = 1/137$,is the electromagnetic coupling constant.

44

To obtain the masses, we need to find the length R_o , theoretically we develop the lengths given by; 1.03741×10^{-39} m, 8.3182×10^{-54}m, 9.4334×10^{-54}m,   and 1.2345×10^{-53}m.

Following the given lengths we respectively obtain the masses;

$M_{GUT} = 10^{16}$GeV, $M_W = 80.18$GeV , $M_Z = 90.82$GeV, and $M_H = 119$GeV respectively.

But at $R_B = R_o$, the mass $M_B = 6.661 \times 10^{19}$GeV is obtained. And at $R_o = 2.529 \times 10^{-37}$m, the Planck mass is obtained (that is $M = m_p$).Therefore it is found out that the W and Z boson particles survive in length of 10^{-54}m . The Higgs particle survives to a length greater than that of the W boson $\geq 10^{-53}$m. And finally particles at the grand unified scale will survive at 10^{-39} m.

5.3.3 The big bang acceleration and proton decay

For proton decay the intensity P is used such that at Schwarzschild radius R and Planck   mass scale m_p the life time of the proton as explained by SUSY is seen to agree so well with the equation $T(time) = \alpha^2 \, m^5_p \, R / 4096 \, \pi^3 m_K^4 \, \hbar$ such that at $m_k = 7.96 \times 10^{-29}$GeV, T $= 10^{35}$yrs. We have obtained the lifetime of protons and the mass of a particle produced during the decay process. The mass of the particle obtained is very small and can therefore be taken to be a neutrino.

The force F_3 can be expressed in the form,

$$F_3 = a_3(m_3^5 / 16\pi^2 m_p^2)^{1/3}$$

Where a_3 is the acceleration, this acceleration at a Planck scale will be given by

$$a_3 = (c^{11} / \hbar\, G^2 m)^{1/3} = 2.4772 \times 10^{52} \, m/s^2$$

This is quite a very large acceleration and therefore defined as the acceleration of particles during the early formation of the universe.

5.4 Discussion

The results obtained describe super symmetry which is a theory required for the unification of everything we know about the physical world into a theory of everything. Significantly a larger enterprise of the theory is to produce a theory of quantum gravity which is required for the unification of general relativity with the standard model, which explains the other three basic forces in physics (electromagnetism, the strong interaction, and the weak interaction), and provides a palette of fundamental particles upon which all four forces act. Theoretically the results obtained (Eqn11 and Eqn12) show a huge correction to the particles' masses, which without fine-tuning will make them much larger than they are in nature. The problem of the unification of the weak interactions, the strong interactions and electromagnetism is solved mathematically, through the comparisons of the cube of the forces in a ratio that generates the gravitational coupling constant power equation.

The Planck mass is the mass of a black hole whose Schwarzschild radius multiplied by π equals its Compton wavelength. The radius of such a black hole is roughly the Planck length, which is believed to be the length scale at which both general relativity and quantum mechanics simultaneously become important. In accordance with the results obtained it is seen that the Planck mass is the mass at which the four forces (F_1, F_2, F_3 and F_p) are equal, the forces are then taken to be related to the origin of the universe simply because at those high energies that formed the dense soup of the universe the forces were equal and the masses probing the Planck mass scale that is black holes were produced, hence those four forces a significant in that they play a crucial role in the formation of black holes. The intensity P on the other hand explains a phenomenon that occurs at the cosmic scale, for example it explains the nature of Black holes and the age of the universe. The acceleration obtained is so large that it is the acceleration that the universe had at the instant after the big bang. Obtaining this acceleration is the possibility of studying the rate of expansion of the universe at large, the accelerating universe is therefore the observation that the universe appears to be expanding at an accelerated rate.

At the Planck scale the descriptions of subatomic particle interactions in terms of quantum field theory breaks down. Also at the same scale, the strength of gravity is expected to become comparable to the other forces, mathematically all the fundamental forces are unified at that scale. The results obtained explain both the weak and strong interactions that at a length between 10^{-37}m and 10^{-35} the Planck scale is attained also at lengths 10^{-39}m , the grand unified scale becomes relevant , but for lengths 10^{-53}m and

10^{-54}m, the standard model holds on well. We have therefore attained a unification that increases from about 10^{-59}m (standard model) to 10^{-35}m (quantum gravity).The paper there fore gives out the relationship between elementary particle physics and astrophysics at a large scale.

5.5 Conclusion

Basing on the results obtained, it is now clearly justified that gravity can be integrated with quantum mechanics at the Planck scale. And therefore the success of the "standard model" which includes both the electroweak theory and quantum chromodynamics can now be regarded as successful in providing accurate descriptions of the fundamental particles and their interactions.

CHAPTER SIX

6.1 On the generalization of loop quantum gravity

Quantum gravity is the field of theoretical physics that tries to unify quantum mechanics with general relativity. Quantum mechanics describes the three fundamental forces of nature while general relativity is a theory of the fourth fundamental force: gravity. The goal every one is waiting for to emerge from this unification is a "theory of everything", or "Grand Unified Theory" (GUT). So many researches have been conducted in line with the theory, for example in 1986, Abhay Ashtekar reformulated Einstein's field equations of general relativity using what have come to be known as Ashtekar variables, a particular flavor of Einstein-Cartan theory with a complex connection. He was able to quantize gravity using gauge field theory. In the Ashtekar formulation, the fundamental objects are a rule for parallel transport and a coordinate frame known as a vierbein at each point. Because the Ashtekar formulation was background-independent, it was possible to use Wilson loops as the basis for a nonperturbative quantization of gravity. Explicit (spatial) diffeomorphism invariance of the vacuum state plays an essential role in the regularization of the Wilson loop states.

Around 1990, Carlo Rovelli and Lee Smolin obtained an explicit basis of states of quantum geometry, which turned out to be labelled by Penrose's spin networks. In this context, spin networks arose as a generalization of Wilson loops necessary to deal with mutually intersecting loops. Mathematically, spin networks are related to group representation theory and can be used to construct knot invariants such as the Jones polynomial.

Quantum field theory depends on particle fields embedded in the flat space-time of special relativity. General relativity models gravity as a curvature within space-time that changes as a gravitational mass moves. Historically, the most obvious way of combining the two (such as treating gravity as simply another particle field) ran quickly into what is known as the renormalization problem. In the old-fashioned understanding of renormalization, gravity particles would attract each other and adding together all of the interactions results in many infinite values which cannot easily be cancelled out mathematically to yield sensible, finite results. This is in contrast with quantum electrodynamics where, while the series still don't converge, the interactions sometimes evaluate to infinite results, but those are few enough in number to be removable via renormalization. Being closely related to topological quantum field theory and group representation theory, LQG is mostly established at the level of rigour of mathematical physics. While confirming that quantum mechanics and gravity are indeed consistent at reasonable energies, this way of thinking makes clear that near or above the fundamental cutoff of our effective quantum theory of gravity a new model of nature will be needed. That is, in the modern way of thinking, the problem of combining quantum mechanics and gravity becomes an issue only at very high energies, and may well require a totally new kind of model.

The general approach taken in deriving a theory of quantum gravity that is valid at even the highest energy scales is to assume that the underlying theory will be simple and elegant and then to look at current theories for symmetries and hints for how to combine them elegantly into an overarching theory. One problem with this approach is that it is not known if quantum gravity will be a simple and elegant theory (that resolves the conundrum of Special and General Relativity with regard to the uniformity of acceleration and gravity, in the former case and space time curvature in the latter case).

6.2 Objectives

The need for the paper is to understand those problems involving the combination of very large mass or energy and very small dimensions of space, such as the behavior of black holes, and the origin of the universe.

The aim of the paper is to answer questions like; How can the theory of quantum mechanics be merged with the theory of general relativity and remain correct at microscopic length scales? What verifiable predictions does any theory of quantum gravity make?

To study the level at which wave mechanics becomes inapplicable, and there fore explain the nature of condensed matter physics using a simple model.

6.3 Materials and methods

6.3.1 The formula for the quantization of quantum gravity

The model is based on separating the gravitational field into the sum of two components; that is the background and the quantum field. The background left is one for all our calculations. But because loop gravity ignores the back ground space as a lost entity that does not occur in space, there fore the need to reconstruct quantum field theory from scratch without a background space is taken into account. I therefore suggest that the calculation should be performed by summing all possible space-times.

Quantum field theory depends on particle fields embedded in the flat space-time of special relativity. General relativity models gravity as a curvature within space-time that changes as a gravitational mass (m) moves.Assuming a spherical symmetric object that space time is of dimensions increasing from 1, 2, 3, 4...N, where N is the nth term of the dimensions. To quantize space and time is to create a space in which all of physics is quantized. The nature of the curved space surface is described by increasing powers in the Schwarzschild radius $R_s = Gm/c^2$, Hence describing the dimensions of space. Quantum mechanics explains the existence of discrete energy states in an atom, in away that the angular momentum of the atom must be quantized, which is also the case for quantum gravity. The equation for the quantization of the loop quantum gravity can then be written as,

$$\eta R_s + \beta R_s^2 + \mu R_s^4 + \ldots\ldots\ldots + \delta R_s^N = n\hbar \qquad [1]$$

Where $\eta = \sqrt{Beh}$, is the momentum of a particle probing another form of quantum mechanics, $\hbar = h/2\pi$, where h is Planck constant, $\beta = 8\pi Be$, e is the elementary charge, B is the magnetic field and finally $\mu = 256\pi^3 P/c^2$, where P is the intensity and c is the constant speed of light.

6.3.2 The energy equation

What changes is the form of the equation the rest remaining constant. The principle behind this is that eqn1 can be changed to any form simply for purposes of calculating complex phenomenon. The energy to which we are concerned here is expressed as a general expression describing the energy scales forming smaller and larger matter entities in the universe. The energy will thus be given by;

$$\eta c + \beta c R_s + \mu c R_s^3 + \ldots\ldots\ldots + \delta c R_s^{N-1} = n\hbar c/R_s \qquad [2]$$

Note: the background space described by the Schwarzschild radius has changed, thus the above equation in any case can be used to calculate the basic properties of Black holes. Remember the Schwarzschild radius is the radius for a given mass where, if that mass could be compressed to fit within that radius, no known force or degeneracy pressure could stop it from continuing to collapse into a gravitational singularity.

6.3.3 The mass equation

Having explored the energy scale we now form general equation that describes well the mass scale. This is also done the same way as eqn2 and therefore generate,

$$\eta/c + \beta R_s/c + \mu R_s^3/c + \ldots\ldots\ldots + \delta R_s^{N-1}/c = n\hbar/cR_s \qquad [3]$$

6.4 Results

6.4.1 The Planck scale and the gravity magnetic field

Assuming that the energy $W = \beta c R_s$, from eqn2 is equal to the energy $W = mc^2$, we hence obtain the magnetic field as, $B = c^3/8\pi Ge = 1.0054 \times 10^{53}$ N/Am. using this magnetic field in the energy equation, $W = \eta c$ we get the energy in the form $W = (c^2/2)\sqrt{\hbar c/G}$ where the quantity $\sqrt{\hbar c/G}$ is the Planck mass M_p at an energy of 6.119×10^{18} GeV.

6.4.2 Time taken by a black hole to evaporate and its entropy

The energy required here is given in Eqn2, it is at this, that the intensity P = W/AΔt, (where A is the area and t is the time) is used. We take the energy $W = \mu c R_s^3$ (from Eq2) as our interest from which we obtain the time as $\Delta t = 256\pi^3 R_s^3/Ac$. But with black holes the area will become exactly equal to the square of the Planck length as $A \sim L_p^2 = \hbar G/8\pi c^3$ hence the change in time is given by $\Delta t = 63500.86\pi G^3 m^3/\hbar c^4$.

For entropy we set the energy to kT, where k is Stefan's-Boltzmann's constant and T is the temperature of the body. Now for $kT = \mu c R_s^3$, since Δt is known the entropy is thus given by $S = W/T = 78.96 Ak c^3/\pi \hbar G \sim A/4$. In conclusion we state that the entropy of a black hole is proportional to the area of the event horizon.

6.4.3 The quantum Hall Effect

For this effect the momentum η is used. From Eqn2 we set, $\eta c = nh / R_s$ which gives the magnetic flux as $4 \pi R_s^2 B = nh/e$, from which the resistance is given by $\zeta = 4 \pi R_s^2 B /e = nh/e^2$. for n= 1,2,3,4 the resistance is of a value 25833.8Ω.

6.4.4 Symmetry breaking at the Planck scale

Using eqn3 in this case, since B is known and P got from μR_s^4 $=nh$; as $P = \hbar c^2/256\pi^3 R_s^4$, we hence obtain, $M_p /2 + m + M_p /m$

$= M_p /m$, which gives $M_p + 2m = 0$, and for identical mass $M = 0$, which is true. The intensity at the planck length that is for $R_s = L_p$ is

$$P = c^8/\pi\hbar\, G^2$$

6.5 Discussion

The results obtained signify the comparisons between quantum mechanics and gravity that at the known Planck scale both theories are combined into one theory. The deduction of black hole thermodynamics implies that the theory is at its best a true quantum gravity theory. The Planck scale is what it is, not because it expresses natural units but because of the magnetic field of about 10^{53} N/Am through which matter interacts with energy to produce particles probing the Planck scale. The success of this theory implies the production of features resembling the standard model, for this I will regard my self to only condensed matter physics as discussed below;

For Compton and photoelectric effect to occur particles with momentum $p = h/\lambda$, which describe the wave properties of matter at a wave length λ, are required but for the quantum Hall Effect to occur the particles must have a momentum $p = \sqrt{Beh}$ at which the magnetic flux and resistance hold on well. This is a darling new idea describing particles probing the wave particle duality model, and therefore addresses a phenomenon referred to as the "exact quantization". The quantum Hall effect is a quantum-mechanical version of the Hall effect, observed in two-dimensional electron systems subjected to low temperatures and strong magnetic fields, in which the Hall conductance takes on quantized values. The momentum η on the other hand indicates a move from a probabilistic theory of quantum mechanics to a complete theory of mechanics. For example comparing this momentum with that of De Brogile you find out that the magnetic flux is in terms of wavelength and given by B λ^2 =h/e, this is known as flux quantization.Flux quantization occurs in Type II superconductors subjected to a magnetic field. The possibility of obtaining such a result from simply combining the different momentums must have an implication. The uncertainties brought by the quantum theory can be foregone by introducing in "quantum mechanics" a new mechanics that states that the outcome of an ideal measurement of a system is deterministic. Therefore the momentum is taken as the separating line between quantum mechanics and condensed matter physics.

6.6 Conclusion

The recent observation of naked singularities and doubly special relativity as part of loop quantum cosmology show a clear future for the significance of the theory to all of physics researchers, but as for now there is no experimental observation to show whether loop quantum gravity makes the predictions not made by the standard model or general relativity however the

successful prediction and calculation of the Bekenstein –
Hawking formula S=A/4 in this paper is evidence to show that
loop quantum gravity is a theory that must be generalized
regardless of it's past mistakes.

Successfully I have made a new mechanics that is different from
that of deBrogile but makes observable results when combined
with the deBrogile hypothesis. Thus a mechanics for condensed
matter has been finally developed.

CHAPTER SEVEN

7.1 The earliest period of time in the history of the universe

This paper represents a brief history of the universe, that is from its past to present. Observations have suggested that the universe began 13.7billion years ago. The universe was so hot with particles having a very high energy, in its earlier phase. The evolution then proceeded with this energy forming the first protons, electrons and neutrons, then nuclei and finally atoms. The microwave background was also emitted during the formation of the neutral hydrogen. Finally the structure of the universe was formed when matters aggregated into the first stars and quasars and on large scale clusters of galaxies and super clusters were formed.

In cosmology, the Planck epoch , named after Max Planck, is the earliest period of time in the history of the universe, from zero to approximately 10^{-43} seconds, it is at this time that quantum effects of gravity were significant. At this period approximately 1.37×10^{10} years ago all fundamental forces were unified. The state of the universe during the Planck epoch was unstable, tending to evolve and giving rise to the familiar manifestations of the fundamental forces through a process known as symmetry breaking. It is currently believed that the Planck epoch inaugurated the Grand unification epoch, and that symmetry breaking quickly led to the era of cosmic inflation, the Inflationary epoch, during which the universe greatly expanded in scale over a very short period of time.

The age of the universe, in Big Bang cosmology, refers to the time elapsed between the Big Bang and the present day. Current observations suggest that this is about 13.7 billion years, with an uncertainty of about +/-200 million years. Extrapolation of the expansion of the universe backwards in time using general relativity yields an infinite density and temperature at a finite time in the past. This singularity signals the breakdown of general relativity. How closely we can extrapolate towards the singularity is debated—certainly not earlier than the Planck epoch. The early hot, dense phase is itself referred to as "the Big Bang", and is considered the "birth" of our universe. Based on measurements of the expansion using Type Ia supernovae, measurements of temperature fluctuations in the cosmic microwave background, and measurements of the correlation function of galaxies, the universe has a calculated age of 13.7 ± 0.2 billion years.[21] The agreement of these three independent measurements strongly supports the ΛCDM model that describes in detail the contents of the universe.

The aim of this paper is to examine the Planck epoch and the grand unification epoch and therefore find out the true scale that explains the earliest period of the universe.

7.2 Methodology

I construct a mathematical model to study the relationship between the ratio of the wavelength and the ratio of the accelerations. In the model I refer to different particles probing different scales of time, I refer to the Planck scale and the quantum gravity scale. Assuming two particles, one under the influence of gravity and the other the influence of quantum fields, when the two particles are set to fall through the fields each falls with an acceleration describing the fields that is g_Q and g_G respectively. The length scales describing the falls are λ_e and λ_c, so the equation below will fully describe the model,

$$\lambda_e/\lambda_c = (g_Q/g_G)^{1/2} \qquad\qquad [1]$$

where, $\lambda_e = e/2\lambda_p E\varepsilon$,is the wavelength of an electron depending on the wave length of a photon λ_p. The higher the photons wavelength the smaller the electrons wavelength and the smaller the wavelength of the photon the higher that of an electron, this takes place on the assumption that the electric field E, the charge e, and the permittivity of free space ε, are all constants. $\lambda_c = \hbar/mc$, is the Compton wavelength of an electron with mass m , here $\hbar$ is Dirac constant and c is a constant speed of light. $g_Q = e^2 f/2\,\hbar\varepsilon$ describes the acceleration of a particle with a frequency f in the quantum field and , $g_G = Gm/R^2$ is the acceleration due to gravity and G is the universal gravitational constant.

for ,$\lambda_p = c/f = 2\pi R$, $E = e/4\pi\varepsilon R^2$ where R is the distance between any two particles.

We compare the forces that come as a result of the motion. There are two forces each occurring at a different scale length. First there is a force describing the Inflation, baryogenesis at the grand unification transition. Second there is another force describing the quantum gravity barrier at the Planck epoch. Both of these forces will have a similar characteristic, which is they will depend on the time t at any level.

Therefore from eqn1 we multiply through by Gc^5 and obtain a relationship of the forces given by,

$$F=t_g(c^7/32\pi mG^2)=t_u(c^7/16\pi G^2m) \qquad [2]$$

Where

$t_u = e^2\hbar/2\varepsilon Gm^3c^2$ and $t_g = A/cR$, A is the area.

7.3 Results

At Planck scale mass $m_p = \sqrt{\hbar c/G}$, $t_u = (e^2/\varepsilon)\sqrt{G/\hbar c^7} = 4.932\times10^{-45}$ s and at the Planck length scale $L_p = R = (\sqrt{\hbar G/8\pi c^3})$, $A \sim L^2_p = \hbar G/8\pi c^3$ and $t_g = \sqrt{\hbar G/c^5}$, then equating t_u to t_g the fine structure constant is generated as $(e^2/16\pi^2\varepsilon\hbar c)$.

7.4 Discussion

The results obtained show exactly the required mechanism responsible for the description of the time line of the big bang.

The reduction of the fine structure constant from the theory shows that there exists a relationship between quantum mechanics and gravity, that at the two times t_u and t_g the force of gravity was strong and that there was a possibility for the unification of all the fundamental forces of nature. It can now be theorized that there exists a scale that when merged with the Planck scale the result is the earliest period of the universe at which all of physics problems can be solved. Therefore both scales are needed to explain the origin of the universe from the big bang.

7.5 Conclusion

Ignoring quantum effects means that the universe starts from a singularity with an infinite density. This hypothesis however can change when quantum gravity is taken into account. The works of String theory , Loop quantum gravity, Noncom mutative geometry and other fields of physics holds promise for our understanding of the very beginning.

However; the more we understand about how matter forms, the more precisely we will be able to interpret what we learn from astrophysical data, and from other sources.

CHAPTER EIGHT

8.1 Another form of general relativity and its new predictions

The development of general relativity followed a publication of acceleration under special relativity in 1907 by Albert Einstein. In his article he argued that any mass will "Distort" the region of space around it so that all freely moving objects will follow the same curved paths curving toward the mass producing the distortions. The questions raised by the principle of equivalence and general relativity are intimately related to the questions of the origin, size, and structure of the universe. Is the universe infinite or finite? How old is our solar system and galaxy? How were they formed? How many other galaxies are there and how are they distributed? Where did they come from? What was the universe like before these galaxies were formed? The field of physics that deals with these questions is called cosmology, a very fast moving field.

In 1916, Schwarzschild found a solution to the Einstein field equations, laying the groundwork for the description of gravitational collapse and, eventually, black holes. In 1917, Einstein tried to describe a static universe, where he added cosmological constant to his original field equations for that purpose. With Hubble's observations in 1929, on the movement of galaxies which predicted an expanding universe, Lemaître formulated the earliest version of the big bang models.

General relativity uses a complex mathematical equation that makes it so hard for people to master the theory. This paper gives out a simple and accurate mathematical formulation of space and time is some what a similar fashion to that of general relativity. This paper deviates from the theory in that for it takes into account the description of space and time for both small (quantum effects) and large particles. This paper also explains the features of cosmology (black holes) and the Big bang (the earliest period of the universe).

Finally, there have been various attempts through the years to find modifications to general relativity. The most famous of these are the Brans-Dicke theory and Rosen's bimetric theory. Both of these proposed changes to the field equations, and both suffer from these changes permitting the presence of bipolar gravitational radiation. As a result, Rosen's original theory has been refuted by observations of binary pulsars. As for Brans-Dicke the amount by which it can differ from general relativity has been severely constrained by these observations. It is generally held that one of the most important unsolved problems in modern physics is the problem of obtaining the true quantum theory of gravitation, that is, the theory chosen by nature, one that will work at all energies. Discarded attempts at obtaining such theories include supergravity, a field theory which unifies general relativity with supersymmetry. In the second superstring revolution, supergravity has come back into fashion, with its as yet undefined quantum completion rebranded with a new name: M-theory.

8.2 Materials and methods

8.2.1 The movement of a particle in a curved path and their associated forces

Since gravity increases in inverse proportion to volume, any quantity of matter that is sufficiently compressed will become a black hole. When a large enough amount of mass is present within a sufficiently small region of space, all paths through space are warped inwards towards the center of the volume, forcing all matter and radiation to fall inward. I formulated a new solution to Einstein field equation which describes black holes, and is given by;

$$\text{Volume} = A_B R_d = (1/ F_e)(h^2/m) \ [1]$$

Where A_B is the area of the small region of space, F_e is the tidal force (An object in any very strong gravitational field feels a tidal force stretching it in the direction of the object generating the gravitational field.) Near black holes, the tidal force is expected to be strong enough to deform any object falling into it, even atoms or composite nucleons; this is called spaghettification. The strength of the tidal force depends on how gravitational attraction changes with distance, rather than on the absolute force being felt. This means that small black holes cause spaghettification while infalling objects are still outside their event horizons, whereas objects falling into large, supermassive black holes may not be deformed or otherwise feel excessively large forces before passing the event horizon.

$R_d = A_e^2 / R_s^3$ is the radius of that region of space, $A_e = hc/ F_e$ is the area occupied by each particle experiencing the tidal force (in other words area of the object). $R_s = Gm/c^2$ is Schwarzschild radius, It is the radius for a given mass where, if that mass could be compressed to fit within that radius, no known force or degeneracy pressure could stop it from continuing to collapse into a gravitational singularity, h is the Planck constant, m is the mass of the object, c is the speed of light and G is the gravitational constant. It should be noted that as the volume R_s^3 increases the radius R_d reduces and as it reduces the radius increases. Therefore R_d is the radius of a region of space that is changed when ever the volume occupied by a compressed mass in that region changes.

Therefore the area is given by,

$$A_B = (R_s^3 / A_e^2 F_e)(h^2/m) = G^3 m^2 F_e / c^8 \qquad [2]$$

The equation obtained shows how the force depends on the area where the mass is concentrated. The above force differs from Newton's gravitational law in that it is directly proportional to the area but inversely proportional to the square of the mass of the body. Hence $Fe = A_B \, c^8 / G^3 m^2 = N \, A_B / m^2$ where $N = c^8 / G^2$

Results

8.3.1 The area gives the forces

Since in Eq2 the force is related to the area we can then use it to obtain the force on any object occupying any given area. If we take a square of the Schwarzschild radius to be the area where if a mass could be compressed to fit within that area, no known force or degeneracy pressure could stop it from continuing to collapse into a gravitational singularity, then the following is obtained

For a black hole of area R_s^2, the force is $F_e = c^4 / G$

For two particles separated by a distance R and within an area R_s^4 / r^2, Newton's gravitational force is $F_e = Gm^2 / r^2$

For particles probing the big bang, the areas are R_s^2 / α_s and R_s^2 / α_g (where $\alpha_g = Gm^2 / \hbar c$ is the gravitational coupling constant and $\alpha_s = ke^2 / \hbar c$ is the fine structure constant, k is coulomb constant and e is elementary charge) the following forces are obtained respectively ,

$$F_G = E_o^2/ke^2 \text{ and } F_E = E_o^2/Gm^2$$

the energy $E_o = \sqrt{\hbar c^5/G}$ where $\hbar$ is the Dirac constant $h/2\pi$. This is the energy describing the scale of the energy that the universe had in its early formation. Therefore substituting the value of F_E in Eq2 we obtain $A_{B1} = \hbar G/c^3 = 2.60624 \times 10^{-70} m^2$ And for $F_G = F_e = E_o^2/ke^2$ we obtain $A_{B2} = A_{B1} (Gm^2/ke^2)$. This means that the force F_E only becomes comparable to F_e at the Planck length scale. And $F_G = F_e$ doesn't achieve the correct scale when the forces are compared, it approaches the scale but with an effect Gm^2/ke^2 which indicates that the gravitational force cannot be compared to the electromagnetic force.

This therefore states that F_G is the gravitational force and F_E is the electromagnetic force at the big bang scale.

8.3.2 The cosmological pressure and temperature
From Eq2 the pressure P is simply a tidal force on an object per unit area occupied by the object in a region of space, hence,

$$P = F_e/A_B = c^8/G^3 m^2 \qquad [3]$$

And finally the temperature from Eq1 is

$$T = (R_d F_e/k) = (1/A_B)(h^2/mk) \qquad [4]$$

69

Where k is gas constant per molecule in joules per Kelvin

8.3.3 The entropy of a black hole and the first law of thermodynamics

8.3.3.1 Entropy of a black hole

Keeping the volume constant, the pressure of a gas is directly proportional to its absolute temperature that is $P \propto T$, hence from Eq3 and Eq4

$$P/T = (\beta^2 A_B / R_s)(k/h^2) \qquad [5]$$

Where β is the rate of change of mass equal to c^3/G

For an ideal gas, keeping the temperature constant, the volume of a gas will vary inversely proportional to its pressure that is $V \propto 1/P$, hence

$$PR^3_s/T = (\beta^2 A_B R^2_s)(k/h^2) \qquad [6]$$

The above equation is the entropy of a black hole derived from the properties of a gas and therefore it can be expressed as

$$\text{Entropy} = P\, R^3_s/T = (kA_B/4A_{B1})(2\beta\, R^2_s/\pi\, h\,) = (kA_B/4A_{B1})\alpha_g$$

$$[7]$$

Where $\alpha_g = (2\beta \, R^2_s \, / \pi \, h \,)$ is the gravitational coupling constant.

8.3.3.2 The first law of thermodynamics

The sum of the kinetic energy and potential energy of all the individual particles making up the system is the internal energy given by

$$U = \Delta Q + \Delta W$$

Where ΔQ is the heat flow into the system and ΔW is the work done by the system. Basing on the results obtained

$$\Delta Q = \sqrt{\hbar c^5 / G} = c\sqrt{\beta \hbar} = E_o \text{ and } \Delta W = -(\beta^2 \, A_B \, R^2_s)(kT/h^2\,) = - P \, R^3_s.$$

Hence the internal energy is formulated as

$$U = (\beta \hbar / E_o)(c^2 - R^2_s E_o[kT/h^2]\,) \quad [8]$$

Letting $[kT/h^2] = 4\pi^2 / \hbar \, \tau$ where τ is the time

We obtain

$$U = (\beta \hbar / E_o)(c^2 - R^2_s E_o[4\pi^2 / \hbar \, \tau]\,) \quad [9]$$

as a result when $U = 0$, $R_s = 1.61414 \times 10^{-35}$ m, and $E_o = 1.9605 \times 10^9$ J. the time

$\tau = 2.1238 \times 10^{-42}$ s, which is the earliest period of the universe is obtained.

Still from Eq8 we find that the quantity $(\beta\hbar/ E_o)$ represents mass which is given by $M_p = 2.1765 \times 10^{-8}$ kg. Multiplying this mass throughout we generate a principle equation

$$U = M_p c^2 - M_p (R_s^2 [E_o kT / h^2]) \qquad [10]$$

This equation gives us a mechanism of combining the laws governing small particles (quantum mechanics) with those governing heavenly bodies (General relativity). The appearance of the Schwarzschild radius R_s which explains galactic bodies, the appearance of the random energy kT that describes small particles, the appearance of the Planck mass M_p and energy E_o which describe the Planck epoch in the early universe are all evidence of the generalized formulation of the combined theory of quantum and gravity hence obtaining a quantum gravity theory of nature.

From which we get the speed of particles in the early universe given by

$$\upsilon =\sqrt{} (R_s^2 [E_o kT / h^2])$$

For $E_o = kT$ the speed is got as $\upsilon = R_s E_o / h = 0.4773 \times 10^8$ m/s which is smaller than the speed of light by only $(\upsilon/c = 0.1591)$

8.3 Discussion

From the results obtained it is studied that the forces acting on heavenly objects depend on the areas of space in which these objects occupy. These areas are also related to the "Schwarzschild area" R_s^2. any area in space will have this effect and it will only change when R_s^2 is divided by a dimensionless constants, for example, the Newtonian gravity will depend on the dimensionless constant R_s^2 / r^2 , the cosmological features depend on a unity dimensionless constant and For forces describing the big bang the dimensionless constants will be the coupling constants determining the strength of the gravitational and electromagnetic fields. When all these dimensionless constants are equal to unity it means that the area occupied by one object in the universe corresponds directly to that occupied by other objects and that the effect of the force to one object is the same to all other objects, therefore implying that the forces will then be unified into one fundamental force.

Forces probing the big bang are directly related to the square of the Planck energy, and it is known that at the Planck scale the description of subatomic particle interactions in terms of quantum field theory breaks down, but since both forces have energies at the Planck scale it means that the two are comparable to the other forces and when they are equated the result is a dimensionless constant which is unity $Gm^2/ ke^2 = 1$.

The thermodynamic laws that describe gases here on earth are seen to be the same laws that govern the particles found in our galaxy. It is seen that Boyle's and Charles laws can be applied to heavenly bodies, the result of this is that the entropy of these bodies is directly proportional to the product of the area of the event horizon of the body and the gravitational constant that determines the strength of the gravitational force.

The earliest period or time line of the big bang is studied. The random energy of particles kT forming matter during that time was in equilibrium with the energy of the photons, the time when this happened was 2.1238×10^{-42} s, every particle during this time moved at a speed of light $c = 3 \times 10^8$ m/s, particles moving at a speed closer to that of light where produced when kT was equal to $E_o = 1.9605 \times 10^9$ J and when calculated had a speed $\upsilon = R_s E_o / h = 0.4773 \times 10^8$ m/s which varies directly with the Schwarzschild radius.

8.4 Conclusion

Successfully I have analyzed a method of combining elementary particle physics with astrophysics. It is now possible to apply the laws governing small particles in the description of the nature of large particles hence the possibility of combining quantum mechanics with general relativity has been given out in detail. The equation for the first law of thermodynamics has also been generalized to $U = M_p c^2 - M_p (R_s^2 [E_o kT / h^2]) = M_p c^2 - M_p (R^2_s E_o [4 \pi^2 / \hbar \tau])$ where τ defines the life time and $M_p = 2.1765 \times 10^{-8}$ Kg is the Planck mass. These results therefore show a clear future for the formulation of the unified law of all of physics.

CHAPTER NINE

9.1 On the complete theory of light

Basing our study on the electric currents generated whenever there is a changing magnetic field (B) and a changing electric field (E) in the electromagnetic wave we can construct a complete theory for the electromagnetic radiations. The theory is created using the symmetry between a long wire placed in the electromagnetic fields which induce vibrating electrons that carry current in the wire and the electromagnetic wave which constitute changing electric and magnetic fields that create vibrating photons in the wave. Therefore a wire is to a wave what a vibrating electron is to a vibrating photon in the wire and a wave respectively. The aim of the paper is to give a clear description of the theory of electromagnetic radiations (light). The goal of the paper on the other hand is to show that the wave-particle descriptions of reality can be applied to any physical situation simultaneously. The objective of the paper is to show that the Photoelectric Effect and the Compton Effect can both be explained by the wave model and the particle model at the same time.

Consider a long wire connected to an ammeter and strong electric and magnetic fields produced in a vacuum. Let us assume that whenever a wire is brought in vicinity of a changing electric field, electrons of mass (m) are set into motion in the wire and then an ammeter deflects, recording a current (i_E). The current in the wire due to a changing electric field should be given by

$$i_E = \frac{j\varepsilon_0}{2\pi m}E \qquad (1)$$

Where (ε_0) is the permittivity of free space and (j) is the constant of action in SI units Js. therefore the current is quantized and depends on both the electric field and the mass of an electron.

When the wire is brought into the magnetic field, vibrating electrons at a frequency of oscillation (f) are set in motion at a speed (v) through the wire generating a current given by

$$i_B = \frac{v}{2\pi \mu_0 f}B \qquad (2)$$

Where (μ_0) is the permeability of free space.

Assuming that the ammeter records different values of (i_E) and (i_B), what will be the change in the current values recorded at the ammeter? Subtracting equation (1) from equation (2) we have

$$\Delta I = (i_E - i_B) = \left(\frac{j\varepsilon_0}{2\pi m}E - \frac{v}{2\pi \mu_0 f}B\right) \qquad (3)$$

This is the change in the currents due to changing magnetic and electric fields. Assuming that there is no change in the current, meaning that the current values for i_E are equal to those of i_B (i.e $\Delta I = 0$). This will imply that the magnetic field strength was equal to the electric field strength at one point in both experiments. In terms of electromagnetic radiations in the vacuum, assuming that a wire carrying current is replaced by a wave and electrons are replaced by photons. The wire replaced by a wave is made up of vibrating electric and magnetic fields at a given frequency making an electromagnetic wave. The electrons replaced with photons will represent the particle properties of the electromagnetic wave (light) with associated mass and speed (v).

The symmetry here is between the long wire and the wave, the electrons and the Photons. The electric and magnetic fields brought in vicinity of the wire and the number of oscillations per second of the electron in the wire is what leads to an electromagnetic wave. The electrons with a given mass and moving at a given speed is what constitute a photon. Then at $\Delta I = 0$, we have on arranging,

$$\frac{jf}{mv} = \frac{1}{2\pi\mu_0\varepsilon_0}\frac{B}{E} \tag{4}$$

This means that at $\Delta I = 0$, either a changing magnetic field or a changing electric field produces a current. Then it should be true that a changing magnetic field produces an electric field just as a changing electric field produces a magnetic field. This process in the electromagnetic wave continues indefinitely. The electromagnetic wave will move at a constant speed (c), since for electromagnetic waves, $\frac{E}{B} = c$, and for a photon $\frac{jf}{mv} = c$ where j=6.63×10^{-34}Js (also called the Planck constant after Max Planck) and mv is the photon momentum. Implying that the photon energy is related to the frequency of the electromagnetic wave by (jf). Then the electromagnetic wave will move at a constant speed given as, since by symmetry $\frac{E}{B} = \frac{jf}{mv} = c$

$$c = \frac{1}{\sqrt{\varepsilon_o \mu_o}} = 2.99792458 \times 10^8 \, \frac{m}{s}$$

Where $\qquad \varepsilon_o = 8.85418782 \times 10^{-12} \frac{c^2}{Nm^2}$ **and**

$\mu_o = 1.26 \times 10^{-6} \frac{Ns^2}{c^2}$

We have therefore deduced based on the symmetry between a current (electron) carrying wire in the electromagnetic field and the photons in electromagnetic waves that an electromagnetic wave moves at a constant speed of light. It is also true from the deductions that light is indeed made up of particles of light called photons and vibrating electric and magnetic fields. The deduction would not be possible if the wave and particle descriptions of the situations had not been applied simultaneously (into what is called "the wave-particle duality).

Unexpectedly enough the **photoelectric effect** can also be explained by Equation (3), on arranging

$$\frac{2\pi m f}{\varepsilon_0 E}\Delta I = jf - \frac{mv}{2\pi\mu_0\varepsilon_0}\frac{B}{E}$$

Then the total energy of the particle of light (Photon) is then given by

$$jf = \frac{2\pi m f}{\varepsilon_0 E}\Delta I + \frac{mv}{2\pi\mu_0\varepsilon_0}\frac{B}{E} \tag{5}$$

It is therefore true that the photoelectric effect can be explained when both the particle and wave models of reality are applied in the experiment at the same time (simultaneously). The work function from Einstein's photoelectric equation (A. Einstein, 1905) will here be replaced by $\frac{2\pi m f}{\varepsilon_0 E}\Delta I$ while the kinetic energy of the electrons at the surface of the metal will be given by $\frac{mv}{2\pi\mu_0\varepsilon_0}\frac{B}{E}$. Equation (5) reduces to Einstein's Photoelectric effect when, the speed of the electron is $v = \frac{1}{\pi\mu_0\varepsilon_0}\frac{B}{E}$ and the change in current for a complete circuit is $\Delta I = \frac{j\varepsilon_0 E}{2\pi m}$.

The validity of the **Compton Effect** can also be deduced from Equation (3). The current can be taken as the product of the frequency (f) of radiations and the charge (q) on the particle. Then the current due to the electric field is $i_E = q f_1$ and that due to the magnetic field is $i_B = q f_2$. In the case of the Compton

Effect, q is the charge on the free electron while f_1 and f_2 are the frequencies of the incoming photon and outgoing photon after collision with the free electron respectively. Then equation (3) can be written as

$$f_1 - f_2 = \frac{1}{q}\left(\frac{j\varepsilon_0}{2\pi m}E - \frac{v}{2\pi\mu_0 f}B\right) \qquad (6)$$

Since photons move with the speed of light(c) then their frequencies is related to their speed and wavelength by $f = \frac{c}{\lambda}$, then we have

$$\frac{1}{\lambda_1} - \frac{1}{\lambda_2} = \frac{1}{qc}\left(\frac{j\varepsilon_0}{2\pi m}E - \frac{v}{2\pi\mu_0 f}B\right)$$

On arranging to include the charge density of the free electron for electric field lines in an area of $\frac{\lambda_1\lambda_2}{2\pi}$, we obtain

$$\frac{2\pi q}{\lambda_1\lambda_2}(\lambda_2 - \lambda_1) = \frac{j}{mc}\left(\varepsilon_0 E - \frac{mv}{\mu_0 jf}B\right)$$

Where (mc) is the momentum of an electron treated relativistically, on letting the charge density $= \frac{2\pi q}{\lambda_1\lambda_2} = \varepsilon_0 E$, we deduce the change in the wave length of the incoming photon and outgoing photon after collision with the free electron as

$$\Delta\lambda = (\lambda_2 - \lambda_1) = \frac{j}{mc}\left(1 - \frac{mv}{\rho\mu_0 jf}B\right)$$

Since $\rho = \varepsilon_0 E$, we then have

81

$$\Delta\lambda = (\lambda_2 - \lambda_1) = \frac{j}{mc}\left(1 - \frac{\frac{mvB}{\varepsilon_0\mu_0 E}}{jf}\right)$$

Since jf is the energy carried by the photon, and then also $\dfrac{mvB}{\varepsilon_0\mu_0 E}$ is the energy carried by the free electron. Treating the electron relativistically such that for electromagnetic waves moving at a speed (v) relative to the electron moving at a speed of light $c = \dfrac{1}{\sqrt{\varepsilon_0\mu_0}}$, the electric field in the wave will be related to the magnetic field by $Bv = E$. then the energy carried by an electron can be given by mc^2. Then the angle at which the photon is scattered after collision with the free electron will be given by

$$\theta = \cos^{-1}\left(\frac{mvB}{\varepsilon_0\mu_0 E}\right)\Big/ jf$$

(7)

Where mv is the momentum of the photon in the electromagnetic wave consisting of a changing electric field E and magnetic field B both moving at a constant speed of light $c = \dfrac{1}{\sqrt{\varepsilon_0\mu_0}}$. Treating the electron relativistically we have

$$\theta = \cos^{-1}\frac{mc^2}{jf}$$

When the energy carried by the photon is equal to the energy possessed by the electron then $\theta = 0$, meaning that there is or there is no scattering and whatsoever there is no increase in photon wavelength hence $\Delta\lambda = 0$.

A complete theory of light can't fail to explain **the structure of an atom**. I therefore take a complete discussion of what goes on inside an atom only with the help of Bohr's energy levels which he derived using classical mechanics and quantum theory. Let $\Delta f = f_1 - f_2$ be an increase in the frequency of the electromagnetic radiations emitted from an atom. Then squaring both sides of equation (6) and arranging will give

$$4\pi^2 \Delta f^2 = \frac{1}{m^2 q^2}\left(j\varepsilon_o E - \frac{mv}{\mu_o f}B\right)^2$$

$$4\pi^2 m^2 q^2 \Delta f^2 = j^2 \varepsilon_o^2 E^2 - 2\frac{j\varepsilon_o E Bmv}{\mu_o f} + \frac{B^2 m^2 v^2}{\mu_o^2 f^2}$$

Dividing through by $64\pi^4 j^2 \varepsilon_o^2$ and multiplying through by q^2 gives the energy of the atom as on arranging

$$\frac{mq^4}{16\pi^2 \pi^4 j^2 \varepsilon_o^2} = \frac{1}{64\pi^4 m \Delta f^2}\left((Eq)^2 - 2\frac{m(Eq)(Bqv)}{\mu_o \varepsilon_o(jf)} + \frac{(Bqv)^2 m^2}{\mu_o^2 \varepsilon_o^2(jf)^2}\right)$$

The energy of the n-th level is, since the reduced Planck constant is $n\hbar = \frac{nj}{2\pi}$

$$\frac{mq^4}{32\pi^2\pi^4n^2\hbar^2\varepsilon_0^2} = \frac{1}{32\pi^2n^2m\Delta f^2}\left((Eq)^2 - 2\frac{m(Eq)(Bqv)}{\mu_0\varepsilon_0(jf)} + \frac{(Bqv)^2m^2}{\mu_0^2\varepsilon_0^2(jf)^2}\right)$$

The expression on the left hand side of the equation is the quantized energy of an atom (Niels Bohr, 1913) while the right hand side of the equation represents the energy of the atom in terms of the forces associated with it. In the equation we let $H_e = Eq$ be the electric force for a particle moving in the electric field and $H_b = Bqv$, the magnetic force on a particle with charge q moving in the magnetic field. Since the speed of light is $c = \frac{1}{\sqrt{\varepsilon_0\mu_0}}$, then the quantized energy can be given as

$$W_n = \frac{1}{32\pi^2n^2m\Delta f^2}\left(H_e^2 - 2\frac{H_eH_bmc^2}{jf} + \frac{H_b^2(mc^2)^2}{(jf)^2}\right)$$

Then on arranging we obtain

$$W_n = \frac{1}{32\pi^2n^2m\Delta f^2}\left(H_e - \frac{mc^2}{jf}H_b\right)^2 \qquad (8)$$

When the energy of an electron moving at a speed of light in atom is equal to the energy of the emitted photon, then

$$W_n = \frac{1}{32\pi^2n^2m\Delta f^2}(H_e - H_b)^2 = \frac{1}{32\pi^2mn^2}\left(\frac{\Delta H}{\Delta f}\right)^2 \qquad (9)$$

Where $\Delta H = H_e - H_b$ is the difference or change between the electric force and the magnetic force in an atom, when the two forces balance (i.e. $H_e = H_b$), then $W_n = 0$ meaning that the total energy of an atom will cease to exist.

Therefore the total energy of an atom increases with the square of the change in the electric and magnetic forces which govern an electron but falls off as the square of the change in the frequency of the radiation emitted by it.

From equation (8) the ratio of the energy of an electron to that of the photon $\frac{mc^2}{jf}$, is the limit at which if the energies are not equal you will not get a change in the electric and magnetic forces. Treating the ratio as a number $\tau = \frac{mc^2}{jf}$, we get from equation (8)

$$W_n = \frac{1}{32\pi^2 mn^2} \left(\frac{H_e - \tau H_b}{f_1 - f_2} \right)^2 \tag{10}$$

When $\tau = 0$, it means that the relativistic energy (mc^2) of an electron in an atom is zero, and that the total energy of an atom only increases with the electric force on the electron. The relationship (equation 10) is a complete expression for the laws according to which, by the theory here advanced, the structure of an atom should be viewed.

In conclusion, a complete theory of light is only possible if both the wave and particle descriptions of reality are applied to the physical situation at the same time. In discussing Young's double slit experiment for example we should be able with the formulas given above to treat the electromagnetic radiations on both a wave and particle model

CHAPTER TEN

10.1 Equations Rendering Magnetic Monopole Creation in a Field Stronger Than Magnetar Fields

By definition magnetars are neutron stars or soft gamma ray repeaters with very high magnetic field strength. They are formed whenever a massive star dies in what is called a type II supernova explosion resulting into a dense ball of subatomic particles of magnetar and ordinary pulsars. Magnetars are a result of fast spins of new born neutron star creating in a process an intense magnetic field while ordinary pulsars are a result of slow spins of newborn neutron stars creating in a process a magnetic field though strong but less as compared to those produced by magnetars.

The magnetic field of magnetars on a theoretical background cannot be deduced using Newtonian mechanics or Maxwellian electromagnetic theory but rather require a more advanced insight and modification of the quantum electrodynamic theory. Below I present a brief but rather rigorous calculation of the magnetic field of ordinary pulsars at the quantum electrodynamic scale and the ultra strong magnetic field of magnetars at the quantum gravity electrodynamic scale.

10.2 Dimensionless constants of nature

The dimensionless constant or in this case a coupling constant is a number that determines the strength of a force involved in any interaction. The known coupling constants of nature are the electromagnetic coupling constant and the gravitational coupling constant. The two couplings play a very great role in calculating the magnetic field of magnetars that is why I have introduced them here.

The electromagnetic coupling constant α_e is a number that determines the strength of the magnetic force in an interaction involving a constant interchange of photons. On the other hand it is also called a fine structure constant that determines the size of the splitting of the hydrogen spectral lines.

$$\alpha_e = \frac{e^2}{4\pi\varepsilon_0 \hbar c} \qquad\qquad 1)$$

Where e is the elementary charge, $\hbar$ is the reduced Planck constant, c is the constant speed of light and ε_o is the permittivity of free space

The gravitational coupling constant α_g is a number that determines the strength of the gravitational force during an exchange of gravitons between elementary particles.

$$\alpha_g = \frac{Gm^2}{\hbar c} \qquad\qquad 2)$$

Where G is the gravitational constant and m is the mass of the particle.

However the significance of both the electromagnetic and gravitational coupling constants described above can be understood using a coupling α_n that determines the strength of a force whenever an amount of electromagnetic radiations of associated magnetic field strength are emitted when an electron jumps from one level of it's lines of orbit to another in an atom. This coupling must be the same in all interactions involving an exchange of particles. in other words it is the same in subatomic particles and extra galactic particles.

10.3 The magnetic field

Thus the magnetic field strength value obtained for the electron revolving in the magnetic field is given by

$$B = \frac{m^2 c e}{4\pi\varepsilon_0 \hbar^2 \alpha_n}$$ **3)**

❖ For **quantum electrodynamic field strength** I set $\alpha_n = \alpha_e$ from Equation 1) from which I get

$$B = \frac{m^2 c^2}{\hbar e}$$ **4)**

89

$$B = \frac{(9.109E - 31)^2 (3E8)^2}{(1.055E - 34)(1.602E - 19)}$$

$$B = 4.3697E13 Gauss$$

This is the magnetic field strength value of a typical surface, polar magnetic field of radio pulsars.

❖ For ultra strong magnetic field strength **(quantum gravity electrodynamic field strength)** I set $\alpha_n = \alpha_g$ from Equation 2) from which I get,

$$B = \frac{ec^2}{4\pi Gh\varepsilon_0} \qquad\qquad 5)$$

$$B = \frac{(1.602E - 19)(3E8)^2}{4\pi(6.67E - 11)(1.055E - 34)(8.85E - 12)}$$

$$B = 1.8423E56 Gauss$$

Such ultra strong magnetic field strengths are capable of breaking down the vacuum and decay via a quantum mechanical process of magnetic monopole creation.

From Equation 3) a table showing different coupling constants at known values of magnetic field is generated

$$B = \frac{0.032E13}{\alpha_n} Gauss$$

Table 1.

Name	Magnetic Field B Gauss	Coupling(strength) α_n	Remarks
Earths magnetic field	0.6	5.33E11	?
Hand held magnet	100	3.2E9	?
Strong sun spots	4000	0.08E9	?
laboratory	4.5E5	0.71E6	?
Man made fields	1E7	1E6	?
Non neutron stars	1E8	1E5	?
Surface, polar magnetic fields of radio pulsars	4.4E13	7.27E-3	Electromagnetic/fine structure constant
magnetars	8E14	4E-4	?

Ultra strong - magnetars	1.8E56	1.78E-45	Gravitational Coupling constant

10.4 conclusion and discussion

From the above analysis it is clear that there are magnetars in our universe with ultra strong magnetic fields whose strength can only be determined at a gravitational coupling scale as shown in table 1. The research undertaken is open for further analysis into this field of physics as it is relevant for theories dedicated towards the development of a theory of quantum gravity.

CHAPTER ELEVEN

11.1. The possibility for man to stay on an atom

By definition an atom is the smallest unit of an element that retains the chemical properties of that element. An atom has an electron cloud consisting of negatively charged electrons surrounding a dense nucleus.

For so long scientists have been extracting electrons from atoms and bringing them onto the earth. Remember atoms have been into existence onto the earth for many years. To me they seem like small universes within a big universe and a combination of them forms the universe within which we live. There can be ways for us to travel to those atoms and see the life there. You think this is fiction, no you are wrong. Look if the atoms merge to form the universe then the universe will separate into small parts to form the atom and this will reduce our size and shape such that we have the dimensions of the universe to which we can survive as you see it today.

If these electrons have the ability to live in the universe then also there must be a possibility for man to live on an atom. For example consider a person going to stay on an atom, remember the atom is very small and man can't even stand on it that is what you think. You will also think that he can't even see where he will land because he can't see the atom with his naked eyes. But the solution is there for a person to stay on the atom.

- ❖ The first ideological formulation of the possibility is by assuming that we are small towards an atom and an electron is big towards the planet.
- ❖ The second assumption is that gravity exists on an atom
- ❖ The third assumption is that the space surrounding the atom is in a frozen state , that is solid like and man can move on it without floating into space
- ❖ The fourth is that the life on the atom for both plant and man is constant

From the above assumptions, you can now imagine bringing a small particle like an electron from the atom onto the earth. Of course it would be easier when you take the above assumptions to be true.

- ❖ The electron it's self will assume to be big for it to stay on the earth and so it does
- ❖ It will also assume that the space is in a frozen state for it to move on it
- ❖ It will also assume that life is constant and extra

The above assumptions are true only if you set your self into motion to the atom. The atom may be an interesting place for you than our universe. **But where is the way to it?**

The way to the atom is your imagination or simply what you think. See if you imagine that you are a small particle then that what you are but when you think that you are a big particle then the impression is true that you are. To go to the atom takes a few seconds than going to the moon. There is of course no transport but distraction of the body.

Your body will break into small particles where by each will be able to land on the atom and finally they will merge at exactly the same dimension as the atom that will make your size and shape fill comfortable on it.

12.0 Ground energy of a hydrogen atom

❖ THE QUANTUM GRAVITATIONAL FORCE

$$H = \frac{c^4}{G}\, \alpha^{n-1}$$

❖ THE QUANTUM GRAVITATIONAL ENERGY

$$w = \hbar \frac{c^4}{pG}\, \alpha^{n-1} \quad (1)$$

Where, $\hbar$- Reduced Planck constant and p- Relativistic momentum

From General relativity an object collapses into a black hole when its size approaches its gravitational radius given as $\frac{2Gm}{c^2}$ (at n=1) while in quantum mechanics the size of an object is determined by its de Brogile wavelength given as $\lambda = \frac{h}{\pi p}$. Equation (1) can then be expressed in terms of the de Brogile wavelength as $w = \hbar \frac{\lambda \pi c^4}{Gh}\, \alpha^{n-1}$. But when the object's wavelength approaches its gravitational radius the total energy of the object will be given as

$$w = \frac{mc^2}{2}\, \alpha^{n-1} \quad (2)$$

From Equation (2) the ground energy of an Hydrogen atom can be deduced when n=3

$$W = \frac{mc^2}{2} \alpha^2$$

Therefore when three lines are close together in the spectrum of an atom, it means that we are defining the discrete energies at a quantum number of n=1 (Atomic level - Protons, neutrons, and electrons).

13.1 References

1) Abhay Ashtekar, New variables for classical and quantum gravity, Phys. Rev. Lett., 57, 2244-2247, 1986

2) Carlo Rovelli and Lee Smolin, Discreteness of area and volume in quantum gravity, Nucl. Phys., B442 (1995) 593-622, e-print available as gr-qc/9411005

3) Castelvecchi, Davide; Valerie Jamieson (August 12 2006). "You are made of space-time". New Scientist (2564

4) Researchers Look Beyond the Birth of the Universe", Eberly College of Science, 12 May 2006.

5) Smolin, Lee. "The case for background independence". hep-th/0507235

6) http://en.wikipedia.org/wiki/Loop_quantum_gravity

7) C.L.Chin and C.R.Westgate (Editors), The Hall Effect and Its Applications," Plenum Press,New York, 1979, p.535.

8) Eddington, A. S., *The* Internal Constitution of *the* Stars (Cambridge University Press, England,1926), p. 16

9) E. Kolb and M. Turner, *The Early Universe* (Addison-Wesley, Reading, MA,1990).

10) W. Garretson and E. Carlson, Phys. Lett. B 315, 232(1993); H. Goldberg, hep-ph/0003197

11) Eddington, A. S., *The* Internal Constitution of *the* Stars (Cambridge University Press, England,1926), p. 1

12) S.P. Martin, in Perspectives on Supersymmetry , edited by G.L. Kane (World Scientific, Singapore, 1998) pp. 1–98; and a longer archive version in hep-ph/9709356; I.J.R. Aitchison, hep-ph/0505105.

13) Mara Beller, *Quantum Dialogue: The Making of a Revolution.* University of Chicago Press, Chicago, 2001.

14) Bohr, Niels (1958). *Atomic Physics and Human Knowledge.* John Wiley and Sons. OCLC 530611 ASIN B00005VGVF.

15) De Broglie, Louis. *The Revolution in Physics*, Noonday Press, 1953.

16) Einstein, Albert. *Essays in Science*, Philosophical Library, 1934.

17) Feigl, Herbert and May Brodbeck, *Readings in the Philosophy of Science*, Appleton-Century-Crofts, 1953.

18) Feynman, Richard P., *QED: The Strange Theory of Light and Matter*, Princeton University Press, 1985. ISBN 0-691-08388-6

19) Prof. Michael Fowler, *The Bohr Atom*, A series of lectures, 1999, University of Virginia.

20) Heisenberg, Werner. *Physics and Philosophy*, Harper and Brothers, 1958.

21) S Lakshmibala, *"Heisenberg, Matrix Mechanics and the Uncertainty Principle"*, Resonance, Journal of Science Education, Volume 9, Number 8, August 2004.

22) Carl Rod Nave, *Hyperphysics-Quantum Physics*, Department of Physics and Astronomy, Georgia State University, CD 2005.

23) F. David Peat, *"From Certainty to Uncertainty: The Story of Science and Ideas in the Twenty-First Century"*, Joseph Henry Press, 2002.

24) Reichenbach, Hans, *Philosophic Foundations of Quantum Mechanics*, University of California Press, 1944.

25) Schilpp, Paul Arthur, *Albert Einstein: Philosopher-Scientist*, Tudor Publishing Commpany, 1949.

26) *Scientific American Reader*, 1953.

27) Sears, Francis Weston, *Optics*, Addison-Wesley, 1949.

28) Shimony, A. (1983). "(title not given in citation)". *Foundations of Quantum Mechanics in the Light of New Technology (S. Kamefuchi et al., eds.)*: p.225, Tokyo: Japan Physical Society. ; cited in: Popescu, Sandu; Daniel Rohrlich. Action and Passion at a Distance: An Essay in Honor of Professor Abner Shimony (PDF). *arXiv.org*. Retrieved on 2007-01-12.

29) Niels Bohr (1913). "On the Constitution of Atoms and Molecules, Part II Systems Containing Only a Single Nucleus". *Philosophical Magazine* **26**: 476-502.

30) Niels Bohr (1913). "On the Constitution of Atoms and Molecules, Part III". *Philosophical Magazine* **26**: 857-875.

31) Niels Bohr (1914). "The spectra of helium and hydrogen". *Nature* **92**: 231-232.

32) Einstein (1917). "Zum Quantensatz von Sommerfeld und Epstein". *Verhandlungen der Deutschen Physikalischen Gesellschaft* **19**: 82-92. *Reprinted in* The Collected Papers of Albert Einstein, *A. Engel translator, (1997) Princeton University Press, Princeton.* **6** *p.434. (Provides an elegant reformulation of the Bohr-Sommerfeld quantization conditions, as well as an important insight into the quantization of non-integrable (chaotic) dynamical systems.)*

33) en.wikipedia.org/wiki/**Planck_epoch** - 23k - Cached - Similar pages

34) en.wikipedia.org/wiki/Bohr_model - 62k - Cached - Similar pages

35) E. Schrödinger, *Ann. Phys. (Leipzig)* **489** (1926) p.79

36) E. Schrödinger, *Phys. Rev.* **28** (1926) p. 10

37) Morrison, Philp: "The Neutrino, scientific American, Vol 194,no.1 (1956),pp.58-68.

38) R. Haag, J. T. Lopuszanski and M. Sohnius, Nucl. Phys. B88, 257 (1975) S.R. Coleman and J. Mandula, Phys.Rev. 159 (1967) 1251.

39) H.P. Nilles, Phys. Reports 110, 1 (1984).

40) P. Nath, R. Arnowitt, and A.H. Chamseddine, Applied N = 1 Supergravity (World Scientific, Singapore, 1984).

41) S.P. Martin, in Perspectives on Supersymmetry , edited by G.L. Kane (World Scientific, Singapore, 1998) pp. 1–98; and

a longer archive version in hep-ph/9709356; I.J.R. Aitchison, hep-ph/0505105.

42) S. Weinberg, The Quantum Theory of Fields, VolumeIII: Supersymmetry (Cambridge University Press, Cambridge,UK, 2000).

43) E. Witten, Nucl. Phys. B188, 513 (1981).

44) S. Dimopoulos and H. Georgi, Nucl. Phys. B193, 150(1981).

45) N. Sakai, Z. Phys. C11, 153 (1981);R.K. Kaul, Phys. Lett. 109B, 19 (1982).

46) L. Susskind, Phys. Reports 104, 181 (1984).

47) L. Girardello and M. Grisaru, Nucl. Phys. B194, 65(1982); L.J. Hall and L. Randall,

48) Phys. Rev. Lett. 65, 2939(1990);I. Jack and D.R.T. Jones, Phys. Lett. B457, 101 (1999).

49) For a review, see N. Polonsky, Supersymmetry: Structureand phenomena. Extensions of the standard model, Lect.Notes Phys. M68, 1 (2001).

50) G. Bertone, D. Hooper and J. Silk, Phys. Reports 405, 279 (2005).

51) G. Jungman, M. Kamionkowski, and K. Griest, Phys. Reports 267, 195 (1996).

52) V. Agrawal, S.M. Barr, J.F. Donoghue and D. Seckel,Phys. Rev. D57, 5480 (1998).

53) N. Arkani-Hamed and S. Dimopoulos, JHEP 0506, 073(2005); G.F. Giudice and A. Romanino, Nucl. Phys.

B699, 65(2004) [erratum: B706, 65 (2005)]. July 27, 2006 11:28

54) en.wikipedia.org/wiki/Supersymmetry - 52k - Cached - Similar pages

55) en.wikipedia.org/wiki/Grand_unification_theory - 39k - Cached - Similar pages

56) Abhay Ashtekar, New variables for classical and quantum gravity, Phys. Rev. Lett., 57, 2244-2247, 1986

57) Carlo Rovelli and Lee Smolin, Discreteness of area and volume in quantum gravity, Nucl. Phys., B442 (1995) 593-622, e-print available as gr-qc/9411005

58) Castelvecchi, Davide; Valerie Jamieson (August 12 2006). "You are made of space-time". New Scientist (2564

59) Researchers Look Beyond the Birth of the Universe", Eberly College of Science, 12 May 2006.

60) Smolin, Lee. "The case for background independence". hep-th/0507235

61) http://en.wikipedia.org/wiki/Loop_quantum_gravity

62) C.L.Chin and C.R.Westgate (Editors), The Hall Effect and Its Applications," Plenum Press,New York, 1979, p.535.

63) Epstein. M., et al, "Principals and Applications of Hall-Effect Devices", Proceedings of the National Electronics Conference, 1959, Vol.15, p.241.

64) Final Engineering Report on Hall Effect Device Investigation", Device DevelopmentCorporation, Weston 93,

Massachusetts, Contract No. NOBsr-72823, July 1, 1958 toFebruary28, 1959, pp.12-17

65) In cosmology, the **Planck epoch** (or **Planck** era), named after Max **Planck**, is the earliest period of time in the history of   the   universe,   ... en.wikipedia.org/wiki/**Planck_epoch** - 23k - Cached - Similar pages

66) L. Shapiro and J. Sol`a, Phys. Lett. B 530, 10 (2002);

67) E. V.Gorbar and I. L. Shapiro, JHEP 02, 021 (2003); A. M. Pelinson,

68) L. Shapiro, and F. I. Takakura, Nucl. Phys. B 648, 417 (2003).

69) A. Starobinsky, Phys. Lett. B 91, 99 (1980).

70) G. F. R. Ellis, J. Murugan, and C. G. Tsagas, Class. Quant. Grav.21, 233 (2004).

71) H. V. Peiris et al., Astrophys. J. Suppl. 148, 213 (2003).

72) D. N. Spergel et al., astro-ph/0603449.

73) Vilenkin, Phys. Rev. D 32, 2511 (1985).

74) A. Starobinsky, Pis'ma Astron. Zh 9, 579 (1983).

75) A.H. Guth, Phys. Rev. D23, 347 (1981).

76) A.D. Linde, Phys. Lett. B108, 389 (1982); A. Albrecht, P.J. Steinhardt, Phys.Rev. Lett. 48, 1220 (1982).

77) A.D. Linde, Phys Lett. B129, 177 (1983).

78) N. Makino, M. Sasaki, Prog. Theor. Phys. 86, 103 (1991); D. Kaiser, Phys. Rev.D52, 4295 (1995).

79) H. Goldberg, Phys. Rev. Lett. 50, 1419 (1983).

80) E. Kolb and M. Turner, *The Early Universe* (Addison-Wesley, Reading, MA,1990).

81) W. Garretson and E. Carlson, Phys. Lett. B 315, 232(1993); H. Goldberg, hep-ph/0003197.

82) Eddington, A. S., *The* Internal Constitution of *the* Stars (Cambridge University Press, England,1926), p. 16

83) Duncan R .C. & Thompson C., Ap.J.392, L 9 (1992).

84) Thompson , C, Duncan , R .C ., Woods , P., Kouveliotou , C ., Finger , M.H. & van Parad ij s , J .,ApJ, submitted , astro-ph /9908086, (2000).

85) Schwinger , J .,Phys. Rev.73, 416L (1948)